J. G. Baretta-Bekker · E. K. Duursma
B. R. Kuipers   (Eds.)

Encyclopedia of Marine Sciences

With 127 Figures

Springer-Verlag

Berlin Heidelberg New York
London Paris Tokyo
Hong Kong Barcelona
Budapest

Drs. Hanneke (J.) G. Baretta-Bekker
NIOZ, Netherlands Institute for Sea Research
P.O. Box 59, NL-1790 AB Den Burg, TEXEL, The Netherlands

Prof. Dr. Egbert K. Duursma
Res. Les Marguerites, App. 15, 1303, Chemin des Revoires
F-06320 La Turbie, France

Dr. Bouwe R. Kuipers
NIOZ, Netherlands Institute for Sea Research
P.O. Box 59, NL-1790 AB Den Burg, TEXEL, The Netherlands

ISBN 3-540-54501-8 Springer-Verlag Berlin Heidelberg New York
ISBN 0-387-54501-8 Springer-Verlag New York Heidelberg Berlin

Library of Congress Cataloging-in-Publication Data. Encyclopedia of marine sciences /
J.G. Baretta-Bekker, E.K. Duursma, B.R. Kuipers (eds.). p. cm. Includes biblio-
graphical references and index. ISBN 3-540-54501-8. – ISBN 0-387-54501-8 (U.S.)
1. Marine sciences – Encyclopedias. I. Baretta-Bekker, J. G. (Johanna G.) II. Duursma,
E. K. (Egbert K.) III. Kuipers, B. R. (Bouwe R.) GC9.E56 1992 551.46–dc20 92–
2877 CIP

© Springer-Verlag Berlin Heidelberg 1992
Printed in Germany

Typeseting: Data conversion by Triltsch, Würzburg
Printing and binding: Triltsch, Würzburg

32/3145 – 5 4 3 2 1 0 – Printed on acid-free paper

Preface

Marine science today is an area of rapid expansion. Scientific activity is increasing, and a growing number of scientists are involved in one or more of its disciplines. For a thorough understanding of the marine environment, for its exploration, exploitation and management, background knowledge of more than one discipline is required. A clear need therefore exists for a concise multidisciplinary oceanographic encyclopedia.

The initiative for this book was taken by the Netherland Institute of Sea Research, NIOZ, Texel, a multidisciplinary institute, founded in 1876.

The encyclopedia contains some 1850 entries, 210 of which in marine chemistry, 330 in physical oceanography, 350 in marine geology and 940 in marine biology, while the remaining 20 are general terms. Concepts, terminology and methods of the various disciplines are briefly explained and especially in the description of marine processes the book tries to be more than a glossary or dictionary. It tries to provide a succinct overview of the major topics in marine science. Students, teachers, and scientists as well as interested laymen may use it to find an explanation of oceanographic terms outside their own fields. The manuscript was read by a variety of potential users, whose comments were used to improve and clarify the descriptions.

We appreciated the good cooperation with Springer Verlag and we trust this concise encyclopedia will be useful for educational and general purposes. Of course an encyclopedia is never complete. We hope we have made a relevant selection of entries, but will be happy to receive comments and additions for possible future editions.

The Editors

Acknowledgments

We would like to thank all contributors to this encyclopedia. Most of the entries were provided by our colleagues of the Netherlands Institute for Sea Research. A few, however, were contributed by outside specialists, to whom we are especially grateful.

Thanks are due to our "test" readers: postgraduates at NIOZ, who read and commented on parts of the encyclopedia and Ab Dral, who read and commented on the whole mansucript. We acknowledge their comments and have incorporated most of these in the texts. Joke Hart-Stam typed most of the texts. Willem Hart took care of the illustrations.

We used NOTIS-IR, an Information Retrieval program on the institute's Norsk Data Computer for sorting etc. of the entries. Spinger-Verlag was provided with the ASCII text on floppy disk.

The Editors

Editorial Board

Key to Different Fonts

The entries are printed in **bold**.

Terms printed like → ocean denote cross-references to entries that will provide additional information.

Terms in the body of an entry printed like dating are defined within that body. So the definition of the entry:

dating see → absolute age can be found within the body of **absolute age**.

Italic is used for:
– the scientific names of genera and species
– variables in mathematical formulas.

A

abiotic lifeless, a term used for characteristics and elements of the environment with a certain influence on survival or reproduction of organisms that are not alive (→biotic) themselves. Examples are temperature, light, nutrients, etc., also denoted as abiotic factors or substances.

absolute age (geol.) age in years. Age measurement is usually based on radioactive dating, a method in which the known rate of decay of a radioactive "parent" element into a stable or radioactive "daughter" element is applied to the measured, present-day ratio of these two elements. There are several useful combinations of elements, such as uranium-lead, potassium-argon (K-Ar), rubidium-strontium, and ^{14}C-^{12}C. The →^{14}C method can be used on carbonate or organic material for direct dating of young sediments, while the K-Ar method is used on →glauconite, a mineral that is formed on the sea floor. See also →radio isotopes.

absorption (1) (chem.) see →sorption; (2) (radiation) see →optics.

abundance (syn. density, biol.) number of organisms per unit of habitat space. Indices of relative abundance may be useful, for example number of birds seen per hour or percentages of samples occupied by a species. See also →concentration.

abyssal hill low relief feature of the ocean floor, ranging up to several hundred meters in height and several kilometers in diameter. About 85 % of the Pacific Ocean floor and 50 % of the Atlantic Ocean floor are covered by abyssal hills. See →sea-floor topography, →seamount and →tablemount.

abyssal plain area of the ocean-basin floor that is flat, with a slope of less than 1 : 1000. They are among the flattest portions of the earth's surface. They occur adjacent to the outer margins of the →continental slope, between 3000 and 6000 m deep. The horizontal extent of an abyssal plain ranges from less than 200 km to more than 2000 km. Some have a flat surface with a continuous gradient in one direction, whereas others are marked by broad irregularities. They are widespread in the Atlantic and Indian oceans and occur in marginal seas such as the western Mediterranean, Gulf of Mexico, and the Caribbean.

abyssal zone zone over the →abyssal plain, see →zonation.

Acanthocephala phylum of parasitic roundworms. The juveniles live in →crustaceans, the adults in the intestines of fish. Eggs develop inside the fish, where the provisional hooks of the →pharynx are formed. Larvae develop in the water; there they are eaten by crustaceans, in which they become mature. Once the host is eaten by a fish, the parasites attach to the wall of its digestive tract.

Acoela order of the class →Turbellaria (phylum →Platyhelminthes or flatworms). Small marine worms.

Acoelomata animal groups which lack a body cavity (→coelom) clearly as a secondary loss. Examples are the phyla →Platyhelminthes and →Nemertea.

acoustic release device used for recovering anchored instruments. It is placed between a submerged buoy and its anchoring and is released from the anchor after receiving an acoustic signal. The buoy with instruments and the release will float to the surface to be recovered, while the disposable anchoring weight stays on the bottom. Also used for opening and closing RMT nets (see →nets).

acoustics science that studies the properties of sound. Sound propagation properties of seawater are important both for many marine animals as a means for sensing and communication, and for navigation, underwater detection, and research. Many developments in the field have a naval background. The sound propagation depends on the variation of sound velocity with temperature, pressure and, to a lesser degree, salinity. The distribution of temperature and salinity in the sea may, under given circumstances, result in →refraction of sound and propagation patterns that may either concentrate sound energy in certain zones ("sound channels") or that may leave other areas outside the area where the sound can penetrate (shadow zone). Acoustic signals may be obscured by background noise (ambient noise); this can be noise produced by organisms, noise from rain or waves, and the noise produced by ships. See also →echosounding and →sonar.

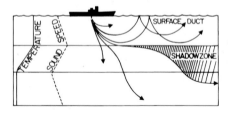

Acoustics.
(After Williams 1973)

acoustic tags see →tagging.

Actiniaria sea anemones, order of subclass Zoantharia, class →Anthozoa (phylum →Cnidaria). Sessile and solitary, usually with basal disk, no skeleton. Cylindrical polyp with a muscular body wall.

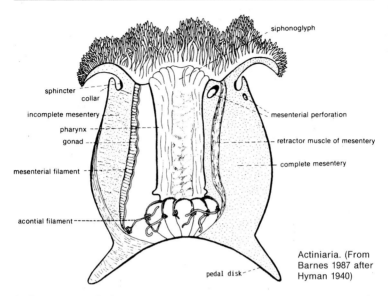

siphonoglyph

sphincter
collar
incomplete mesentery
pharynx
gonad

mesenterial filament

acontial filament

mesenterial perforation

retractor muscle of mesentery

complete mesentery

pedal disk

Actiniaria. (From
Barnes 1987 after
Hyman 1940)

Actinozoa see →Anthozoa.

action spectra see →light.

active continental margin see →continental margin.

active transport process the mechanism by which living organisms are
capable of transporting solutes against a concentration gradient. Active
transport counteracting physical diffusion requires energy. It enables an
organism to maintain a lower concentration of sodium inside the cells and it
enables the cells to accumulate certain nutrients inside the cells at
concentrations much higher than extracellular concentrations. The exact
mechanism of active transport is still unknown, although many explanations have been proposed. Possibly a carrier molecule is involved which
reacts chemically with the solute to be actively transported. The compound
formed is soluble in the lipid portion of the membrane barrier and moves
through the membrane, against the existing concentration gradient, to the
other side, where the solute is released. The carrier molecule then diffuses
back to pick up another molecule. Carrier molecules are probably proteins;
each molecule to be transported requires a specific carrier.

activity coefficient factor which, when multiplied by the molecular or ionic
concentration, yields the chemically active mass of substance in solution. It
is the correction factor which adapts concentrations for thermodynamic
calculations. The activity coefficient is obtained from physicochemical data.

adenosine triphosphate (ATP) chemical compound involved in energy-requiring reactions of cellular metabolism in living organisms. It contains three phosphate groups bound to adenosine; two of them, the so-called high energy phosphate bonds, yield a very high energy (7 kcal/mol) on hydrolysis. This chemical energy can drive an energy-requiring reaction. It can also be used to activate a compound prior to its entry into a particular reaction. The activation by ATP permits a thermodynamically unfavored metabolic process to occur. Some ATP is produced by reactions that occur in the cytoplasm, but about 95% of all ATP is produced within the →mitochondria, organelles concerned principally with the generation of energy. The energy-rich phosphate groups are produced at the energy cost of, for instance, the complete oxidation of glucose.

adiabatic process a process is adiabatic when the system does not exchange heat with the surroundings. According to the second law of →thermodynamics the work done in such a process by compression or expansion causes a change in the →enthalpy. Adiabatic lifting of a volume of seawater to the sea surface gives a decrease in enthalpy, and thus cooling. The resulting temperature in such a thought-experiment is called the →potential temperature, symbol θ, and this parameter is used to characterize →water masses that may move vertically, as it is independent of such motions (conservative parameter).

adsorption see →sorption.

advection transport by the mean current, opposed to →eddy diffusion or spreading by →dispersion.

aeolian sediment see →eolian sediment.

aerobic organisms organisms that require free oxygen for their metabolic processes; they may be facultative or obligate aerobic.

aerobic processes processes, which can only be carried out in the presence of free oxygen.

aerobic zone oxygen-containing zone in the sea bottom, see →benthic system.

age determination of organisms (biol.) determination of the age of organisms is a requirement, for instance, in studies of production and population dynamics. Length may give an indication, but hard parts may show (daily) growth-lines and (annual) winter rings, which allow more precise age determinations. Such growth lines occur in →otoliths and scales of fish, in molluskan shells, in teeth of whales, etc. See also →absolute age.

age distribution the relative frequency of individuals of different age classes within a population, see also →population parameters.

aggregation process resulting in uniting single particles into larger units (aggregates, flocs). In the sea this occurs primarily by organic matter that glues mineral particles together. Colloid particles are also flocculated by electrolytes dissolved in the water (salt flocculation). Suspended particles can be brought together by turbulence, Brownian movement (only particles $< 1-8$ μm), by differences in settling rate or by organisms bringing particles together in feces, or pseudofeces. Small particles are easily attached to larger ones settling through the water column (scavenging). In the sea large flocs $(0.1-2.0$ mm diameter), called →marine snow, are ubiquitous, but fragile so that they are broken during sampling. They can only be studied by →in situ observation (diving, underwater photography). See also →flocculation.

air bladder see →gas bladder.

air gun see →seismic instruments.

air-sea exchange (or ocean-atmosphere exchange) The physical and chemical properties of the ocean and the atmosphere are influenced by their mutual exchange. (1) (phys.) The exchange of heat (including →latent heat), water, and momentum at the sea surface, between the air and the sea strongly influences the physical conditions in both the ocean and the atmosphere. This exchange is studied by measurement of the transport processes in the boundary layers and by assessing the budgets of heat, water, and momentum in these two interacting media. For practical purposes formulas have been developed that allow for the computation of this exchange from observed values such as wind, temperature, etc. The →turbulence in the boundary layers is an important factor in the exchange. The exchange of momentum is expressed by the →wind stress. As the exchange of water by →evaporation also involves the latent heat, the heat exchange is also affected. The global differences in the heat and water exchange between ocean and atmosphere are an important factor in the climate system of the earth. The ocean circulation that transports heat from the equator to the poles has a considerable effect on the moderation of the →climate. (2) (chem.) The exchange of chemical substances can be by different ways. Gases, fluid droplets, as well as particles (sea salt, aerosols), can act as the physical state for exchange. The rate of exchange is difficult to quantify but is largely dominated by the wind-driven mixing of the surface waters. At low wind velocity a very thin so-called sea-surface microlayer may control the exchange rate. Gas exchange rates can be assessed from vertical distributions of natural radioisotope ^{222}Rn (radon gas). Important for climatic conditions is the exchange of →carbon dioxide.

albedo percentage of the incoming short-wave radiation that is reflected (direct or diffuse) by a surface. For the sea surface the albedo strongly depends on the angle of incidence of the radiation and the roughness of the sea surface.

Alcyonacea order of subclass Octocorallia, class →Anthozoa (phylum →Cnidaria). Soft corals. The skeleton is internal, consisting of →spiculae in the →mesogloea. Colonies may be mushroom-shaped. Mainly tropical. Example *Alcyonium* (dead man's finger).

Algae (Latin for seaweeds), large group of →autotrophic organisms that range in size from microscopic to several meters long, and live in the sea, in freshwater, in moist surroundings on land, and even inside other organisms (see →symbiosis). The algae comprise the autotrophic part of the →Prokaryotes: the Cyanobacteria (Cyanophyceae or blue-green algae), the autrotrophic part of the kingdom Protista (see →Protozoa): all the unicellular algae except for the blue-green algae, and a part of the kingdom of plants, in which a number of algal classes together with the mosses and fungi form the "lower" plants, the counterpart of the "higher" plants which have distinct leaves, roots, stem, etc. The algae are separated into several divisions, subdivided into classes. This systematic grouping is done according to the →pigments; unicellular and multicellular classes can be found together in the same division. Next to the prokaryotic →Cyanobacteria, there are the →eukaryotic division Rhodophyta with one single class, the →Rhodophyceae or red algae, the division Heterokontophyta with the classes →Chrysophyceae or golden-brown algae, →Xantophyceae or yellow-green algae, →Bacillariophyceae or diatoms, →Phaeophyceae or brown algae, and the small (ten species) freshwater class Chloromonadophyceae. Further there are the division Haptophyta with the single class →Haptophyceae comprising the →Coccolithophorida, the division Eustigmatophyta (a small group newly separated from the Xantophyceae), the division Cryptophyta with the single class →Cryptophyceae, the division Dinophyta with the single class →Dinophyceae or autotrophic dinoflagellates, the division Euglenophyta with the only class →Euglenophyceae. Finally, there is a group of algal classes →Chlorophyceae or green algae, the small class →Prasinophyceae, which is distinguished by small organic scales, and the class →Charophyceae or candleworths.

algal mat ecosystem benthic shallow-water system in which the sediment is overgrown with dense layers of filamentous and thallous algae and their associated microorganisms. The occurrence of algal mats is related with the absence of an active benthic macrofauna with its grazing and →bioturbation, while on the other hand, the dense algal mat prevents benthic animals from settling in the seabed. Estuarine mudflats have been observed to change temporarily into algal mats after →oil spills. See also →benthic system.

alginates additives to food (cream, pudding) to make fluids firmer, originating from the →Phaeophyceae.

aliasing misinterpretation of a recorded signal due to undersampling. Oceanographic variables generally vary in two or three spatial directions and in time. If, for instance, a →plane wave in the propagation direction is

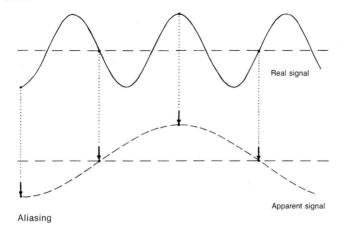

Real signal

Apparent signal

Aliasing

measured, the variance →spectrum will contain variance at much smaller wavelengths (see →wave characteristics) than if the same wave is measured in a direction nearly parallel to the wave crests. Therefore additional assumptions or theoretical or empirical knowledge are required to interpret undersampled recordings of this type. Another type of aliasing can occur in a digitized analog signal. If a time-varying signal is recorded every time interval δt, the highest frequency that can be interpreted with this sampling rate is the Nyquist frequency $1/2\delta t$, corresponding to a wave period of $2\delta t$. Frequencies above the Nyquist frequency are therefore undersampled, although they will contribute to the recorded total variance of the parameter. This contribution may wrongly be attributed to lower frequencies. This type of aliasing can be compared with the effect of the wheels of a motion picture apparently revolving backwards.

alkalinity total (or titration) alkalinity, (A_T), was defined by N.W. Rakestraw as the excess of bases (proton acceptors) over acids (proton donors) in seawater. The expression for total alkalinity is commonly presented as $A_T = [HCO_3^-] + 2[CO_3^{2-}] + [B(OH)_4^-] + [OH^-] - [H^+]$. This expression defines the equivalence point for an alkalinity determination as the point (pH 4.5) at which $[H^+] = [HCO_3^-] + 2[CO_3^{2-}] + [B(OH)_4^-] + [OH^-]$. Total inorganic carbon (C_T), is expressed as $C_T = [H_2CO_3] + [HCO_3^-] + [CO_3^{2-}]$. Both quantities A_T and C_T can be precisely determined from a single, potentiometric titration. They play a fundamental role in the study of the oceanic inorganic carbonate system.

allochthonous from elsewhere, as opposed to →autochthonous: produced →in situ. Used, e.g., for organic matter that may be imported from elsewhere, or for organisms.

allometric growth see →growth.

allopatric speciation see →speciation.

aluminum (Al) is after →oxygen and →silicon the most abundant element (about 8.3 % Al by weight) in the earth's crust. Terrigenous marine sediments consist largely of aluminosilicates, among which the various clay minerals produced through chemical weathering. In seawater dissolved Al exists mostly as the hydrolyzed $Al(OH)_3$ and $Al(OH)_4^-$, whereas speculation exists on finely dispersed colloidal forms. Dissolved concentrations range from 0.5 nmol l^{-1} or less in deep waters of the central North Pacific to about 150 nmol l^{-1} in deep Mediterranean waters. The latter waters exhibit a direct relationship between dissolved Al and dissolved Si, whereas in the Atlantic and Pacific Oceans there appears to be an inverse relation between dissolved Al and Si. Dissolved Al in seawater appears to be controlled by a combination of biological and nonbiological processes.

ambergris see →whale fisheries.

ambulacra radial bands of tube feet projecting from the surface of →Echinodermata.

ambulacral groove groove in →Echinodermata in which rows of tube feet are situated.

amino acid white, crystalline organic compound containing a basic amino (NH_2) group attached to the same carbon atom that holds the acidic carboxyl (COOH) group, to be isolated from →protein hydrolyzates. Amino acids may be classified according to the number of their amino and carboxyl groups: (1) acidic amino acids, containing an excess of carboxyl groups (aspartic and glutamic acid); (2) basic amino acids, containing an excess of basic nitrogen (e.g., histidine, arginine, lysine); (3) neutral amino acids, containing one amino and one carboxyl group, which may be subdivided according to whether the radical of the restgroup in the general formula represents an aliphatic (e.g., glycine, serine, leucine), an aromatic (phenylalanine, tyrosine), a sulfur (cysteine, methionine, cystine) or heterocyclic (e.g., tryptophan, proline) nucleus. Amino acids are soluble in water, except cystine and tyrosine, and insoluble in alcohol and ether (except proline). They form crystalline salts with metallic bases and with mineral acids. Many taste sweet (glycine, alanine, serine, and proline), others are tasteless (tryptophan and lecrine) or bitter (arginine). Fundamental constituents of living matter; hundreds to thousands of amino acids are

NH2
|
R — C — COOH
|
H   Amino acids

combined to make each protein molecule. There are about 25 amino acids known to be constituents of →proteins, and synthesized by →autotrophic organisms, such as most green plants. Certain "essential" amino acids must be obtained from the environment by →heterotrophic organisms. Other "nonessential" amino acids are not necessarily required from the environment, since organisms can synthesize them themselves.

ammonia gaseous compound (NH_3), which in solution in water is partially present as ammonium ion (NH_4^+). In freshwater systems as well as in the marine environment, it is a basic nutrient for phytoplankton growth. In the decompositon of organic compounds (manure), ammonia and ammonium ions are major end products, which can be transported through the atmosphere over considerable distances. Ammonia can be oxidized microbiologically to nitrate (→nitrification), and thus form an important source of "acid rain". See also →nitrogen cycle.

ammonification is the formation of ammonia from decomposing nitrogenous organic compounds, see also →nitrogen cycle.

ammonites (Ammonoidea) extinct order of →cephalopod mollusks with coiled external shells with complex suture lines. They have been extinct since the →Cretaceous.

Amoebida (amebas, amoebas) order of the subphylum →Sarcodina (phylum Sarcomastigophora, see →Protozoa), which are naked (sometimes enclosed in a protein shell), and live in the sea, freshwater or on land in moisture. The cell changes constantly in shape, and crawls over the substrate with pseudopodia. Some species reach several mm in length. Example: *Amoeba*.

amphidromic point see →co-tidal lines.

Amphineura former class of the phylum →Mollusca, now divided into the classes →Polyplacophora and →Aplacophora.

Amphipoda (amphipods) order of class →Malacostraca, subphylum →Crustacea (phylum →Arthropoda). Large order (100 families) of small and mainly marine, but also freshwater crustaceans. There is no →carapace, only the first two segments are fused to the head. The abdomen is not distinctly demarcated from the thorax, and there are seven pairs of legs.

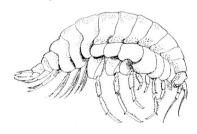

Amphipoda

They resemble the →Isopoda, but the body is usually laterally compressed. The compound eyes are sessile. Example: gammarids.

Amphiuroidea see →Ophiuroidea.

anadromous (upward-running) term used for marine fishes that migrate from the sea to freshwater for their reproduction, e.g., the salmon (opposite of →katadromous).

anaerobic organisms organisms that do not require free oxygen for their metabolic processes.

anaerobic processes are processes which can be carried out, in the absence of free oxygen. See also →anaerobic respiration.

anaerobic respiration the use of an electron acceptor other than oxygen in the electron-type transport oxidation system. This type of respiration occurs in some groups of bacteria. The most common anaerobic electron acceptors are nitrate, sulfate, and carbonate, which are eventually metabolized into ammonia, H_2S, and methane, respectively.

anaerobic zone see →benthic system.

analysis (chemical) determination of chemical constituents by various techniques. In the marine environment, a great number of these techniques have to be adapted due to the excess of sodium chloride in seawater relative to all other major and minor constituents. Marine chemical analyses are used in physical oceanographic research, marine chemistry itself, and marine biology and geology. A requirement is that the analytical instrumentation is adapted to sea-going conditions on board ship.

anemometer instrument for measuring the wind velocity. It may be based upon different principles. The speed sensor is usually mechanical (e.g., the cup-anemometer with a vertical axis), but inertia of mechanical sensors makes them less useful for measuring the short-time fluctuations of the wind. For this purpose other devices (hot-wire, acoustic, or pressure-difference sensors) are in use. The direction of the wind is determined by a wind-vane, or, in the case of the latter type sensors, sometimes by measuring the different components of the wind (N-S, E-W, and up-down) separately.

animal excretion the removal from the body cells and the bloodstream of metabolic wastes. This definition implies that the substances must have taken part in cellular metabolism, like carbon dioxide (the end product of oxidative metabolism) and ammonia (the end product of nitrogen metabolism). The substances should also have no further use: the release of substances from cells which are utilized either locally or elsewhere in some body process is called secretion. Both excretion and secretion involve an expenditure of energy in contrast to defecation, which refers to the elimination of wastes that have not taken part in a cellular metabolic process.

Annelida phylum of segmented worms, including earthworms and many freshwater and marine species. The nervous system, the excretory organs, the vessels, and the →coelom follow the segmentation, but the gut is a straight canal from the mouth to the posterior anus. The phylum Annelida comprises the classes →Polychaeta (many →chetae on the segments), →Oligochaeta (few chetae) and →Hirudinea (leeches).

anodic stripping voltametry electrochemical technique relying on the principle of polarography for direct determination of very low concentrations of →trace metals (Zn, Cd, Pb, Cu) in seawater. By chemical manipulation (pH adjustments, UV irradiation) of the seawater solution one may also obtain insight into the fractionation of a given metal between free ionic state and organically complexed state. Here the difference between measurement of the untreated sample (free metal ion) and the treated sample (total metal content) provides a value for the amount of complexed metal.

anoxic processes or →anaerobic processes. Microbiologically mediated processes, which occur only in the absence of free oxygen.

anoxygenic photosynthesis is the use of light energy to synthesize →ATP by cyclic photophosphorylation without oxygen production, e.g., in green and purple bacteria.

Anthozoa (older name Actinozoa) class of phylum →Cnidaria, comprising sea anemones, stony corals, and octocorals. Sedentary, solitary, or colonial →coelenterates with a →polyp generation only. Polyps are more complexly organized than in other Coelenterata. →Coelenteron divided by at least eight →mesenteries, which bear the gonads, derived from the →endoderm. The Anthozoa are divided into two subclasses. The first subclass is Octocorallia (polyps with eight tentacles) with the orders Stolonifera (organ pipe corals), Telestacea, Alcyonacea (soft corals with example *Alcyonium* or dead man's finger), Helioporacea or Indo-Pacific blue coral, Gorgonacea or horny corals from tropical and subtropical seas, and Pennatulacea or sea pens. The other subclass is the Zoantharia (polyps with more than eight tentacles), with the orders Zoanthidea (without skeleton), →Actiniaria or sea anemones, →Madreporaria or Scleractinia or stony corals, Corallimorpharia or corals without skeleton, →Ceriantharia (anemones buried in sand), and the Antipatharia or black corals.

antibiotics are chemical agents produced by one organism which are harmful to other organisms; e.g., the microorganisms *Streptomyces, Penicillium*, and *Bacillus* produce the antibiotics streptomycin, penicillin, and bacitracin, respectively. Thousands of antibiotic substances are known and yearly hundreds more are discovered. Antibiotics of various kinds have been detected in the sea. These are produced especially by →algae and reach active concentrations, particularly in phytoplankton →blooms. Probably they play a role: at the time of the phytoplankton maximum the saprophytic bacteria have reached their annual minimum.

antibody a protein present in the serum or other body fluid that combines specifically with an →antigen.

anticyclonic flow see →cyclonic flow.

antidune bedform produced by unidirectional flow of water; shape is in phase with standing surface water waves. May move upstream or downstream; name derived from possible upstream movement. See also →sedimentary structures.

antigen substance, usually macromolecular, that induces specific antibody formation in the serum or other body fluids.

antinode see →standing wave.

aphotic zone ocean layer deeper than where enough sunlight for →photosynthesis can penetrate. Begins, depending on the →turbidity, at a depth of several meters (coastal) to several hundred meters (open ocean).

Aplacophora class of the phylum →Mollusca. Small, worm-shaped mollusks (solenogasters). The foot, mantle, and shell characteristic of mollusks are absent. They are found in the oceans down to 9000 m. They live on the bottom, or on →corals, and are collected by dredging on the ocean floor. Their biology is poorly known.

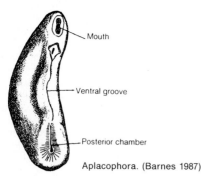

Mouth

Ventral groove

Posterior chamber

Aplacophora. (Barnes 1987)

aposematic coloration see →coloration in marine animals.

Appendicularia see →Larvacea.

aquaculture mass production of aquatic organisms by human effort for commercial purposes. This includes the culture of bacteria, algae, small invertebrates, mollusks, crustaceans, and fish, both in freshwater and in seawater. For the culture of marine organisms in seawater the term mariculture is used. See also →sea farming.

aquatic animals all animals living in or on (fresh- or sea)water. Usually divided into →plankton, →nekton, →benthos, and →neuston.

aquatic plants all plants living in or on water: rooted in the bottom or free-floating.

aragonite see →carbonate, mineral.

Archaebacteria an evolutionarily distinct group of →prokaryotes, including the methanogenic, extremely →halophilic and sulfur-dependent bacteria.

Archiannelida formerly separate phylum, or class of the phylum →Annelida, now regarded as a variety of modified →Polychaeta, mainly belonging to the →meiofaunal, hence microscopic, worms with simplified structure, usually without →setae. Nervous system unsegmented, in or directly under the epidermis. Example: *Polygordius.*

archipelago sea area that contains numerous islands, or the island group itself.

areal distribution horizontal distribution of organisms over the earth; area occupied.

arrow worm see →Chaetognatha.

Arthropoda (arthropods) phylum of segmented animals of which the body is entirely covered with a chitinous exoskeleton or →cuticula. This phylum comprises, with at least three quarters of a million species, more than three times the total of all other existing species. Arthropoda are divided into the subphyla →Trilobita (fossil), Chelicerata (classes Merostomata or horseshoe crabs: *Limulus* sp.; Arachnida or spiders, scorpions, ticks, and mites; Pycnogonida or seaspiders), →Crustacea and Unirama (classes: Insecta, Chilopoda, Diplopoda, Symphyla, and Pauropoda).

Aschelminthes group of small animals (microscopic to several millimeters), marine and freshwater, that used to be regarded as one phylum: Aschelminthes, but that at present is divided into the separate phyla: →Gastrotricha, →Nematoda, →Rotifera, Nematomorpha (parasitic in →Arthropoda), →Acanthocephala, →Kinorhyncha, Loricifera (discovered in 1983, microscopic, interstitial in marine gravel), →Priapulida and, Gnathostomulida (microscopic, interstitial in →anaerobic mud).

Ascidiacea class of the subphylum →Urochordata (phylum →Chordata), sessile tunicates, common all over the world on solid shallow sea bottom,

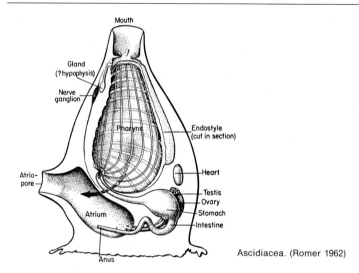

Ascidiacea. (Romer 1962)

rocks, shells, constructions; with two openings on top: the buccal inhalant →siphon, and the atrial siphon.

assimilation incorporation of digested food and other materials, after absorption, into living protoplasm. Includes CO_2-assimilation in plants, the uptake of CO_2 from water or air, during →photosynthesis.

assimilation efficiency ratio between utilized energy and total energy received, e.g., the portion of light fixed by plants of the total amount of light passing the vegetation, or the portion of food metabolized (and thus not egested undigested) of the total food taken up. Low (ca. 1%) in light fixation by plants. Increases in animals with decreasing ash-content of the food: in detritus-feeders ca. 40%, higher in suspension-feeding zooplankton (60–95%), highest in animals or bacteria feeding on high energy food like sugars or aminoacids (max. 100%).

assimilation ratio (syn. assimilation number) the rate of primary production per unit →chlorophyll-a and unit of time [g O_2 or g C (g Chl-a)$^{-1}$ h^{-1}].

association (1) group of organisms occurring together because they have similar environmental requirements or tolerances; (2) group of minerals occurring together in a rock, esp. in sedimentary rocks.

Asteroidea subclass of the class Stelleroidea (phylum →Echinodermata). Free-moving, star-shaped echinoderms, with arms not sharply marked off

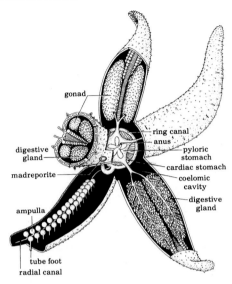

Dissection of a sea star (dorsal view)   Asteroidea. (Keeton 1972)

from the central disk. They have a →madreporite, open →ambulacral grooves with rows of tube feet ending in suckers. The alimentary canal extends into the arms. Examples: *Asterias rubens, Astropecten irregularis, Crossaster papponis.* Starfish or sea stars are the counterpart of the other subclass of the Stelleroidea, the →Ophiuroidea or brittle-stars.

asthenosphere see →earth's internal structure.

astronomical tides astronomical variations in the mutual positions of earth and moon and earth and sun cause variations of the →tides by variation of the tide-generating potential, at roughly monthly, annual, and longer periods. As a result, the tides can be considered the combination of a number of so-called astronomical tides that can be divided into a group of semi-diurnal tides (period about 12 h), diurnal tides (period about 24 h) and longer-period (fortnightly, monthly, semi-annual, annual, and even longer-period) tides. See →harmonic analysis of tides and →partial tides.

Atlantis mythical land, originally mentioned by Plato (429–348 B.C.) and located west of the Atlas Mountains, one of the Pillars of Hercules (the ancient name of the Strait of Gibraltar). It was an archipelago larger than North Africa and Asia Minor (present-day Turkey) together, and flooded by the sea, the present Atlantic Ocean. By some it is considered to have been a real archipelago once and various regions have been proposed as the original Atlantis, in western Europe (north Sweden, Holland, Heligoland) as well as in the Mediterranean, in the Sahara and in Nigeria. Recently it was shown that the present island of Thira in the Aegean Sea was in antiquity a much larger (volcanic) island that may have been Atlantis. The "lost land" of Atlantis for many centuries has appealed to fantasy, and most probably it was only that.

atmosphere the atmosphere that surrounds the earth consists of a mixture of gases of which nitrogen (about 78 % in volume) and oxygen (about 21 %) are the main constituents. Important for the role of the atmosphere in the physical and biological conditions on earth are the content of →carbon dioxide and water vapor. The atmospheric pressure at sea level may vary because of differences in thickness and temperature of the overlying air mass. The horizontal pressure differences exert forces on the air that give rise to →winds; the patterns of the wind and of atmospheric pressure are related in a similar way as in the →geostrophic balance in the sea.

atmospheric pressure see →atmosphere.

atoll see →reef.

atomic absorption spectrophotometry (AAS) technique for the determination of metals. The principle is based on the fact that an atom in its "ground state" is capable of absorbing energy. If energy of the right magnitude (specific wavelength) is applied to an atom, the amount of absorbed energy gives an estimate of the amount of the element present. The atom cloud required for atomic absorption measurements can be produced by supplying enough thermal energy to the sample to produce free atoms. This can be done by aspirating a sample solution into a flame which is aligned in the light (energy) beam, or by means of a graphite furnace, in which the atom cloud is produced by electrical heating of the graphite tube. By the latter method the atom cloud stays inside the tube, which for most elements gives a higher sensitivity and so a lower detection limit than the flame method.

ATP see →adenosine triphosphatem.

attenuation see →optics.

Aufwuchs see →periphyton.

autecology study of relationships between the organism and its environment, on the level of the single species. See also →ecology.

authigenesis general term for the processes that form new mineral phases on the seabed, either by direct precipitation from seawater or by the alteration of pre-existing material. The term literally means "self-originating" and it was originally applied at a time when it was thought that the chemical reactions were entirely inorganic, that is, they occurred without the intervention of marine organisms. The name has stuck, and it is important to emphasize that the term is now applied to reactions that occur essentially in the sediment or at the interface between seawater and the seabed. However, it also encompasses the inorganic precipitation of mineral phases in the water column, for example, the formation of nonbiogenic →calcium carbonate in certain areas (e.g., the warm shallow waters of the Bahamas Banks), and the formation of particulate manganesium oxide in the dispersed plumes from →hydrothermal vents.

autochthonous produced at the spot (→in situ) used for organic matter, but also for animals and plants which (almost) complete their life cycle in a given area (opposite of →allochthonous).

autolysis the self-dissolution which tissues undergo after death of their cells, due to the action of their own enzymes.

autotroph see →metabolic diversity.

auxotrophy the need for a special growth factor. Especially in the →bio-assay technique, auxotrophic microorganisms are used to measure the concentration of the compound the organisms need.

avicularia highly specialized individuals in colonies of →Bryozoa which resemble a bird's head. They prevent larvae and other small animals from settling upon, and interfering with the feeding activity of, the colony.

axenic pure, uncontaminated. See →pure culture.

B

Bacillariophyceae (diatoms), class of the division →Heterokontophyta, unicellular algae with cell walls of amorphous silica, formed by two halves that close together as lid and box, called the frustules. The cells range from 2 μm to a few mm. Some of them form chains of up to 1 cm or more. Diatoms contain the pigments →chlorophyll-c and →fucoxanthin, and store carbohydrate as fluid chrysolaminarin. They can be divided into mainly →benthic, elliptic so-called pennate diatoms, and usually →planktonic centric diatoms which have radial-symmetric frustules. As a group, diatoms are adapted to higher nutrient supplies than other algae, and dominate the →phytoplankton in →upwelling regions. *Chaetoceros, Nitzschia, Rhizosolenia*, and *Thalassiosira* are planktonic genera. *Navicula* is a benthic genus. *Navicula* and relatives move on the substrate by means of mucilage extrusion. This contributes to the consolidation of young sediments. The rich fossil record of diatoms starts in the Jurassic; they are used extensively for →biostratigraphy and →paleoecology.

back-arc basin see →marginal basin.

background radiation see →radiation, background.

backwash see →wave approach.

Bacteria (bacteria) uni-cellular organisms, that are →heterotrophic, and form together with the →autotrophic →Cyanobacteria the prokaryotes (see →prokaryotic). Their size varies from cells as small as 0.1 μm in width to more than 5 μm in diameter. Several distinct shapes of bacteria can be recognized. A spherical or egg-shaped cell is called a coccus, a cylinder shape is called a bacillus or a rod, a curved cell is a spirillum. Coiled bacteria are spirochetes and there are also bacteria with stalks, the appendaged bacteria. Sometimes the cells remain together in groups or clusters or in long chains. Some bacteria can form →spores. Growth occurs by cell division. Like the diversity in forms, the metabolic types of bacteria appear infinite, filling virtually every environmental →niche. Active metabolizing bacteria are found in ice →brines of the polar seas at temperatures below zero as well as in the hot water of several hundred degrees above zero in the deep-sea hydrothermal vents. They can tolerate hydrostatic pressures of the deep sea and grow under all kinds of oxygen concentrations. They are present in anaerobic sediments and anaerobic basins. Bacteria obtain their energy from the oxidation (→aerobic or →anaerobic) of organic material or of reduced inorganic material or from solar energy through bacterial →photosynthesis. Bacteria play a major role in the recycling of elements, the bioenergetics of →ecosystems, the degradation of organic compounds, oil degradation, degradation of →xenobiotics, etc. The role of bacteria as a sink or link of organic matter in the sea is the subject of mineralization and microbial-loop studies.

bacteriocin an agent produced by certain bacteria that inhibits or kills closely related bacteria.

bacterioneuston the thin layer of pelagic bacteria in the uppermost surface of the sea.

bacteriophage virus that attacks bacteria.

bacteriostatic capable of inhibiting bacterial growth without killing.

barnacles see →Thoracica and →Cirripedia.

baroclinic flow if in the ocean the vertical density distribution varies from place to place, the →isobaric surfaces are no more everywhere parallel to the sea surface; this in contrast to the →barotropic case. The horizontal pressure gradient in the water now depends on depth, and the →geostrophic balance gives a baroclinic flow that changes with depth, according to the so-called thermal wind relation.

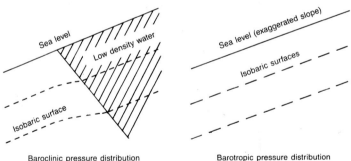

Baroclinic pressure distribution   Barotropic pressure distribution

Baroclinic/barotropic flow

barophile able to live optimally at high hydrostatic pressures.

barotolerant able to tolerate high hydrostatic pressures, although growing better at normal pressures.

barotropic flow in the ocean the →isobaric surfaces are parallel to the sea surface if the density of the water is constant, or when the vertical density distribution in the water is everywhere the same. This situation is called barotropic, and the flow that in such a situation is derived from the →geostrophic balance is independent of depth.

barrier island long, narrow, sandy island situated above high tide and parallel to the shore, usually with dunes and vegetation areas. Also a detached portion of a barrier beach between two inlets. See also →coast.

barrier reef see →reef.

basal metabolism the rate of energy expenditure of an organism at rest, expressed per unit of weight. Measured by calorimetry or by measuring oxygen uptake or carbon dioxide production.

basalt a fine-grained volcanic rock dominated by dark-colored minerals, consisting of plagioclase feldspars (more than 50 %) and ferromagnesian silicates. The typical rock of oceanic crust, generated at →mid-ocean ridges and →rifts. See also →volcanism.

base-exchange capacity in clay →clay minerals have →ion-exchange properties with respect to a number of cations in solution. The extent of these properties can be characterized by the base-exchange capacity. This factor is determined by having all exchange sites filled by Na^+, followed by a replacement with NH_4^+. The amount of Na^+ thus obtained is a measure [in mEq $(100$ g dry material$)^{-1}$] for the base-exchange capacity.

baseline see →coastline, legal aspects.

bathyal zone see →zonation.

bathymetry (the study of) depth distribution, where depth is the distance between seabed and mean →sea level.

bathypelagic zone see →zonation.

bathyscaphe literally "depth ship" a gasoline-filled "underwater balloon" with a gondola for passengers that is used to descend to the deep ocean floor, in 1948 designed by Auguste Piccard. The bathyscaphe Trieste II descended in January 1960 to the bottom of the →Mariana trench.

bathythermograph instrument that records the variation of temperature with depth. These instruments have been devised to be launched from a ship underway, so that in a rather short time, by successive launches, a picture can be obtained of the thermal structure of the upper layers of the ocean. The first instruments measured temperature and pressure by mechanical means and the record was a trace made by a pen on a small sooted plate of glass.They were dropped from the ship, attached to a freely running wire on which they were hauled aboard again to recover the record. Now they have been replaced by expendable bathythermographs (abbreviation: XBT) that measure temperature electrically and send their signal to the ship via a thin electric wire that breaks when the instrument has reached its maximal depth.

bdellovibrios small vibroid microorganisms which have the unique property of preying on other bacteria, using as nutrients the cytoplasmic constituents of their hosts. These bacterial predators are small, highly motile cells, which stick to the surface of their prey cells.

beach gently sloping zone, typically with a concave profile, extending landward from the low-water line, and consisting of unconsolidated material. Landward it ends at the foot of dunes or cliffs or at a zone of permanent vegetation, which is usually the effective limit of storm waves. During periods of exceptionally high tides and strong winds flooding may extend further inland. A beach consists of sand, gravel (shingle, pebbles) and/or shells but there may be accumulation of other loose water-transported material (mud, rock fragments, boulders, material of organic origin). See also →coast.

beach cusp a morphological sandy structure of the beach that is thought to be a manifestation of the presence of →edge waves.

beam trawl see →trawl.

Beaufort scale for estimating wind force. See →wind.

becquerel (Bq) is the modern notation for an amount of radioactive radiation; 1 Bq = 1 disintegration per second (dps). The notation becquerel has replaced the original notation of →curie, where 1 Bq = $2.7 \ 10^{-11}$ Ci. The unit is named after A.H. Becquerel (1852–1908), who discovered radioactive radiation in 1896.

bed forms see →sedimentary structures.

bedload see →sediment transport.

Beer's law applicable to colorimetric or spectrophotometric determination of dissolved substances. Light irradiated into a system (e.g., surface of water masses) can be reflected, transmitted, scattered, refracted, or absorbed. The amount of light absorbed depends on the thickness of the medium that is traversed and is proportional to the number of absorbing molecules per unit volume. In 1852, Beer described, for many solutions of absorbing compounds, that the linear absorption coefficient (a in the law of absorption: P. Bouguer, 1729, and rediscovered by Lambert) is proportional to the concentration of solute c. Thus, Beer's law is $\log I/I_0 = -ecx$, where c is the molar concentration, I and I_0 are the intensities of the measured and source light, respectively, x is the thickness of the liquid layer and e is the molar extinction coefficient. The absorption laws obey strictly only for monochromatic light.

belemnites (Belemnoidea) extinct order of →cephalopod mollusks with an internal shell. They appeared in the fossil record from the →Carboniferous to the end of the →Cretaceous.

bell jars transparent (if light penetration is necessary) or dark jars that are placed →in situ upside down on the seabed, thus enclosing part of the benthic system plus some overlying water. Used to measure exchanges between sediment and water of, e.g., oxygen, nutrients, organic compounds, etc. In deep areas, bell jars are mounted in heavy frames, while automatic instruments stir the enclosed water and register data provided by electronic sensors. Modern bell jars are fully computerised, and operated with an →acoustic release.

benthic referring to the sea bottom.

benthic boundary layer refers to the (water) layer just above and in the surface of the seabed. The biological, chemical, and physical properties are controlled by the presence of, and activity at, the boundary between the seabed and the overlying water. The benthic boundary layer is of particular interest in studies on microbial activity, sedimentary processes, and transports of organic matter, oxygen, and nutrients.

benthic community respiration measure of aerobic metabolic activity of the complete →benthic ecosystem comprising benthic algae, animals and microorganisms. Estimated from the decrease in oxygen concentration in enclosures containing an undisturbed sediment core with overlying water. Measurements can be made on incubated sediment samples, or →in situ with automatic equipment on the seabed (e.g., in →bell jars). The aim of such measurements is the estimation of the amount of organic matter used by the benthic system for respiration.

benthic system ecosystem at the sea bottom comprising the sediment with its associated plant and animal life. This includes the upper layer of the sediment where dissolved oxygen is present (aerobic zone) and the deeper anoxic layer (anaerobic zone) inhabited by anaerobic bacteria and sometimes temporarily by larger macrofauna. The form and functioning of a benthic system is strongly affected by sediment type and depth. In shallow waters where sufficient light penetrates to the bottom, the benthic system may include primary producers (→algae), whereas in deeper waters the benthic system generally receives its food (algae, →detritus, fecal pellets) from the overlying water. Sediment type largely determines the species composition of the benthic system. Schematically, a benthic system could be regarded as a black box receiving organic matter from the →pelagic and oxygen as input (in shallow systems also light, CO_2 and inorganic nutrients), and with inorganic nutrients and CO_2 as output, together with low amounts of organic matter in the form of benthic food eaten by fish or birds, or pelagic larvae released by the →benthos. The overall functioning of benthic systems in the whole of the marine ecosystem is often investigated with →bell jars.

benthos all organisms living on or in the sea floor, also called benthic organisms. Their substrate is species-specific and may be either soft sediments (sand, mud, peat) or hard rock or wood. Benthic organisms live either on the surface of the sediment (→epibenthos) or buried in the

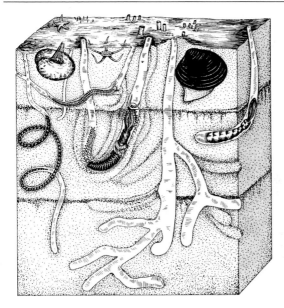

Benthos

sediment (endobenthos). They may be either plants (phytobenthos) or
animals (zoobenthos). Large benthic organisms belong to the →macroben-
thos (or even →megabenthos), smaller ones to the meiobenthos or even
→microbenthos. The boundaries between these groups are not sharp, and
in most species young stages belong to another group than adults.
Macrobenthic animals are big enough to be retained by a 1-mm sieve,
whereas meiobenthos is smaller and separated from the microbenthos by an
arbitrary cutoff point at 100 µm. Benthos is particularly important in the
productive shallow (coastal) water (→benthic systems) where phytobenth-
ic communities receive sufficient light to grow (see →production, primary)
and to feed dense populations of benthic animals. In the sea, food is
produced almost exclusively in the surface water layers. Only small
proportions (usually in the order of a few percent) of the material sink to the
deeper water layers. Hence, →biomass and production of benthic
organisms are low in areas deeper than 1000 or 2000 m. On the other hand,
benthos is often well developed in intertidal and other littoral areas. High
values of biomass are reached particularly in areas like coral reefs, tidal mud
flats, and in restricted bottom areas with high sedimentation in shelf
seas.

Bernoulli's theorem in steady or →irrotational flows there exists a function
of pressure, fluid velocity, and geopotential which is constant along a

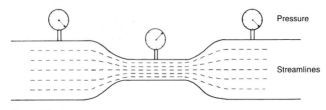

Bernoulli's theorem

streamline. The practical implication is that the pressure drops as the flow velocity increases, as in a contracting canal, and vice versa. This principle, established by the Swiss mathematician D. Bernoulli (1700–1748), is applied in a velocity measuring →pitot tube in which the pressure difference is representative of the current speed.

beta-plane approximation see →Coriolis parameter.

bicarbonate a hydro salt of H_2CO_3, having HCO_3^- as anion. HCO_3^- plays a major buffering role in seawater with the result that the pH of seawater is relatively constant between values of 7.8 and 8.3. HCO_3^- concentrations in seawater are about $18 \ 10^{-4}$ mol at pH 8.3 (see also →alkalinity).

bioassay a technique in which organisms are used as biological sensors in measurements (i.e., on concentration of substances) or the detection of effects of pollutants. Is also the quantitative estimation of biologically active substances by the extent of their actions under standardized conditions on living organisms. Certain organisms can be used as accumulators of chemical compounds, which may otherwise be very difficult to measure, and also to rank toxic pollutants in order of toxicity. An example of a bio-assay is the monitoring of the water quality (→toxicity) of the river Rhine with trout as a "reagent".

biochemical cycles see →biogeochemical cycles.

biocoenose (syn. biocoenosis) community of organisms of different species living together with mutual interactions in an →ecosystem.

biocycles see →biogeochemical cycles.

bioenergetics quantitative description of energy flow or →flux components in organisms, →populations or →ecosystems, usually expressed in joules, kcal or g C m^{-2} y^{-1}, sometimes in terms of P, N, etc. In an energy budget $I = R + \Delta W + U + F + E$, where I is the quantity of food ingested, R is the quantity dissimulated in respiration, ΔW is the gain in body weight, while U, F, and E stand for the quantities lost in urine, feces, and excretion, all per unit of time. Note that ΔW may include reproductive materials.

biofacies see →facies.

biogenic sediment sediment produced directly by the physiological activities of organisms, either plant or animal. Most biogenic sediments consist primarily of the remains of organisms, such as parts of skeletons or tissues of animals. Examples: →coral reefs, →ooze, peat.

biogeochemical cycles more or less closed-circle paths of chemical elements, including all the essential elements for protoplasm passing through from environment to organisms and vice versa. See →carbon cycle, nitrogen cycle, phosphorus cycle and →sulfur cycle.

biogeocoenosis (syn. geobiocoenosis) →ecosystem, a →community interacting with its abiotic environment.

biogeography describes the distribution of organisms on earth and analyzes the causes of the geographical distribution of living and extinct taxa; biogeography of the animal kingdom is called zoogeography. Biological life occupies the whole earth surface but taxa occur clustered in their own habitat and not uniformly distributed. The explanation of discontinuous ranges is based on the assumption that two taxa (species or groups of higher rank) have never developed independently in widely separate areas. The following realms or regions with different assemblages of organisms were previously distinguished: Palearctic (Europa, North Africa and northern Asia), Ethiopian (Africa south of the Sahara), Oriental (tropical Asia and western Indonesia), Australasian (eastern Indonesia, Australia and Polynesia), Nearctic (America north of the tropics) and Neotropical (tropical and southern America). From pole to equator the following provinces in the sea are distinguished: arctic-antarctic, →boreal-antiboreal, warm temperate and tropic, but this division is less marked in pelagic regions than in shelf and littoral regions. Ranges may (have) become discontinuous either by climatic or geological events (→plate tectonics) or by a founder population that has settled beyond the previous border. See also →invasions.

biohorizon [also datum level (less accurate)] isochronous level based on events that can be correlated over a wide area; in a biostratigraphical sense, biohorizons represent evolutionary events such as the first appearance datum (FAD) and the last appearance datum (LAD). Care must be taken to distinguish between a "first appearance datum" and "first occurrence", the latter being the observation of the first appearance within a (local) stratigraphic sequence. Also confusing is the practice (by oil geologists) of using the term FDA for the first downhole appearance (as seen from the top of the section, i.e., the last appearance datum when used in a chronological sense). The (time- or rock-stratigraphic) interval between the first and the last appearance of a taxon is called its range (or stratigraphic range).

biological clock phenomenon present in organisms with an endogenous periodicity, i.e., the →rhythm persists, even when information from the outside world is lacking. Without synchronizing signals (Zeitgebers) from the outside world, however, free-running biological clocks, as experiments with activity rhythms of animals show, get out of phase and run too slowly or too fast. With the aid of their biological clocks animals can perform, e.g., daily or circadian rhythms (circa = about, dies = day) and yearly cycles in their behavior and physiology, or even shorter rhythms like tidal periodicity, while, e.g., swimming activity of small pelagic animals can show cycles of minutes.

biological cycles see →biogeochemical cycles.

biological monitoring see →monitoring.

biological oxygen demand (BOD) yields an estimate of the concentrations of organic matter in the water available to microorganisms. The oxygen uptake is usually calculated by substracting the oxygen content after 5 days' incubation at 20 °C in the dark from the 100 % oxygen content at the start of incubation.

bioluminescence (sometimes wrongly called →phosphorescence) the production of light by living organisms. In the oceans bioluminescence is often caused by →Dinophyceae such as *Noctiluca*. The organisms produce a light flash that lasts about 1/10 s, and the flashes are produced after mechanical, chemical, or other stimulation. Luminescent bacteria often grow on spoiled fish; these bacteria are not →pathogenic. Some fish possess special organs in which luminescent bacteria grow. The bacteria belong mainly to the genus *Photobacterium*. The ability to emit light occurs in many other organisms such as fireflies and fungi. The light is due to an enzyme-catalyzed biochemical reaction. The light-emitting enzyme system of the firefly luciferine-luciferase, is used in the quantification of →adenosine triphosphate (ATP). ATP is the energy source of the light production.

biomass the amount of living material present at a certain moment in a certain area (or within a certain space), expressed in weight units, usually per unit of area or volume, thus $g\ m^{-2}$ or $g\ m^{-3}$. Weight can be given as wet (or live) weight, dry weight (after drying to constant weight) or ash-free dry weight (after subtraction of inorganic remains after ashing). Ash-free dry weight roughly equals weight of organic matter. Carbon weight is ca. 40 % of ash-free dry weight. →Conversion factors between wet and dry weight or between dry and ash-free dry weight are highly variable from species to species and even within species (depending, for instance, on the season). Biomass values are mostly high in coastal areas, particularly on and in the bottom (see →benthos) and low in the oceans.

biometry literally "the measurement of life". From the Greek bios (life) and metron (measure). More generally, biometry is the application of statistical methods to the solution of biological problems. It is also referred to as biological →statistics or simply biostatistics. Adolphe Quetelet (1796–1874), a Belgian astronomer and mathematician, may have been the first biometrist, applying statistical methods to problems of biology and medicine. He introduced the concept of the "average man" and developed the notion of statistical variation and →distribution. Francis Galton (1822–1911), sometimes called the father of biometry, applied statistical methodology to the analysis of biological variation and studied regression and correlation in biological measurements. The largest contribution to biometry has been made by Ronald A. Fisher (1890–1962).

biosphere surface layer of the planet Earth and the atmosphere containing living organisms, a layer a few kilometers thick.

biostratigraphy stratigraphy based on the fossil content of rocks. Biostratigraphical units are called "zones", and are based either on fossil ranges (taxon range zones) or on associations (assemblage zones); see →biohorizon and →stratigraphic hierarchy. The main objective of biostratigraphy is the correlation of stratigraphical sequences in different areas. Major obstacles to this are lateral changes in →facies. See also →stratigraphy and →micropaleontology.

biota living organisms, e.g., freshwater, marine, terrestrial biota.

biotelemetry see →telemetry.

biotic pertaining to life.

biotic environmental factors influences on the structure and functioning of an →ecosystem caused by living organisms, e.g., predation, parasitism, competition for food or space, bioturbation, shading, etc. Opposite of →abiotic factors.

biotin a growth vitamin (H), $C_{10}H_{16}O_3N_2S$, of the vitamin-B complex. Occurs widely in various food products, and is abundant in liver, kidney, yeast, cauliflower, and peas. It is an essential nutrient for various microorganisms and higher animals. Biotin deficiency in animals is associated with the development of dermatitis, loss of fur, disturbances of the nervous system, and finally death.

```
              O
              ||
              C
           /     \
       HN          NH
        |           |
       HC ——————— CH —— (CH2)4 —— COOH
        |           |
      H2C          CH
           \     /
              S                        Biotin
```

biotope part of the biosphere in which the conditions of life for a population of living organisms are approximately homogeneous.

biotoxin →toxin of biological (natural) origin.

bioturbation a term derived from biological perturbation of soil and sediment and applied for activities of benthic organisms (→macro-, meiobenthos) which result in reworked sediments. Burrowing worms, →bivalves and →crustaceans often disturb the sediment structure considerably by mixing and pushing deeper material upwards (diffusive type of bioturbation). Some →deposit-feeding animals have a special grain-size selective effect on the sediment composition by deposition of feces (consisting of sediment eaten under the sediment surface) on the seabed (conveyor-belt type of disturbation). The feeding activities of fish, shrimps, and birds can also cause mixing of surface sediment. The mixing and selection of sediments by bioturbation, together with physical turbation by currents and waves, strongly affect the exchanges of organic matter, oxygen, nutrients, microorganisms, etc. between the →anaerobic zone, the →aerobic zone and the overlying water. Bioturbation is an important item in geological studies on sedimentation rates (see →ichnofossil). Bioturbation strongly affects the chemical composition of pore waters. The reworking under marine conditions may have the qualitative extent of a diffusion process with apparent diffusion coefficients of about 10^{-7} to 10^{-8} cm^2 s^{-1} for sediment reworking and 10^{-4} cm^2 s^{-1} for pore-water mixing. For comparison, molecular diffusion in pore waters is only about $3 \cdot 10^{-6}$ cm^2 s^{-1}.

birth rate see →natality.

Bivalvia (bivalves) class of the phylum →Mollusca. This class is also known as the Pelecypeda or Lamellibranchia, and contains all the common bivalves, like mussels, cockles, clams, etc. The body is bilaterally symmetrical and lies in the mantle cavity, enveloped by an extension of the body wall, the mantle. The mantle is divided into two lobes, excreting the shell valves. The valves are joined dorsally by a ligament and closed ventrally by one or two adductor muscles. The head is poorly developed and bears two →labial palps for the selection of the food material transported to the mouth side of the mantle cavity by the →ctenidia. Ctenidia are often misleadingly referred to as "gills", though the uptake of oxygen in the Bivalvia occurs mainly through the mantle. The water circulation enters and leaves the mantle cavity via often prolonged →siphons. Unused food particles leave the shell through the inhalant siphon as pseudofeces; digested food leaves the gut as feces through the exhalant siphon. The sexes are always separated, eggs are released in the water; the development of the larvae is often →pelagic (see also →meroplankton). (Figure see p. 29)

black shale dark shale, exceptionally rich in carbonaceous organic matter and sulfide. It is formed by the partial anaerobic decay of buried organic matter in quiet-water environments (e.g., →stagnant waters), which are character-

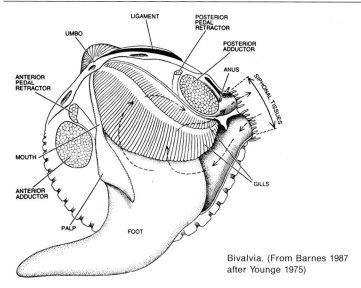

LIGAMENT
POSTERIOR PEDAL RETRACTOR
UMBO
POSTERIOR ADDUCTOR
ANTERIOR PEDAL RETRACTOR
ANUS
SIPHONAL TISSUES
MOUTH
ANTERIOR ADDUCTOR
GILLS
PALP
FOOT

Bivalvia. (From Barnes 1987 after Younge 1975)

ized by very slow deposition of detrital (see →detritus, geol.) materials. See also →deep-sea sediment.

blooms in marine ecosystems in nature, reproduction and density of species are normally limited by the available resources like space, nutrients, and energy, by predation, and other mortalities. When during short periods these limitations are absent, explosive increases in number are possible, called blooms. Good examples are the phytoplankton spring blooms characteristic of the seas of the temperate region: Due to low light irradiance, there is only a very low production of algae during winter, and since growing algae are the main consumers of dissolved inorganic nutrients, N-, P- and Si-concentrations increase. When in spring irradiance exceeds a certain threshold, the production season starts with exponential increase in numbers of diatoms and flagellates. In temperate seas, 50–75 % of the annual →primary production occurs during these spring blooms in March-April. Other examples are →red tides: blooms of flagellates in the surface water layer after exceptional temperature increase in summer.

blue-green algae see →Cyanobacteria.

BOD see →biological oxygen demand.

bongo nets see →nets.

boomer see →seismic instruments.

bore a sharply developed front of a tidal flood wave, occurring in some rivers or estuaries with a large tidal range, and caused by their typical topography. The flood entering over a shallow bottom slope is deformed into a steepening front that travels upstream at the speed of a long wave. The bore marks the transition from subcritical flow downstream to supercritical flow upstream (see →Froude number). The Tsing Kiang river in China has a strong bore with a front up to 4 m high. In England, the Severn estuary has a well-known bore.

boreal areas geographical region with short summer and long cold winters. See also →biogeography.

borers, marine organisms that live in holes that they bore in hard substrates. Borers are found in several taxonomic groups: →Fungi, Algae, Porifera, Bryozoa, Phoronida, →Bivalvia, Sipuncula, →Polychaeta, Crustacea, even in the →Echinodermata and →Cirripedia. Borers usually penetrate substrates by mechanical means: wood, calcareous algae, corals, shells of mollusks or barnacles, sandstone, limestone, chalks, and even bricks, but not granite or concrete. Special measures have to be taken to protect marine cultures and constructions against damage by borers. Shell-boring sponges (*Cliona*) can be a pest in oyster cultures and wood-borers can destroy constructions such as ships and sluices. Even submarine cables need special protection, as some animals bore through lead casing.

bottom friction see →friction.

bottom nepheloid layer in deep ocean waters containing fine-grained sediment, resuspension occurs due to near-bottom currents. The turbid layer of several meters is called bottom nepheloid layer.

bottom samplers variety of scientific equipment used for sampling bottom sediments and benthic organisms. Depending on the aim, and the wide variation in local conditions, sediment type, water depth, and size or lifting gear of the research vessel, bottom samplers range from simple hand-operated corers in →intertidal areas to complicated bottom →trawls, sledges, dredges, grabs, box corers, and deeply penetrating geological corers in →subtidal areas and at the ocean bed. In the early days of oceanographic exploration, most bottom sampling was by dredging (see →dredge), and the primary purpose was to collect marine life. At the beginning of the century the →grab sampler was devised and large sediment samples could be obtained. Such samples, however, are usually somewhat disturbed. If the stratification of the sea floor is of interest, →cores of the sea bottom are preferable to grab samples. The use of →drilling to study the deep ocean bottom was not seriously considered until the 1960s. Ever since, as drilling from an unanchored ship became feasible, deep-sea drilling has proved to be a powerful tool in obtaining scientific information on the sea floor. See also →Deep Sea Drilling Project, →JOIDES, →corer and →box corer.

bottom topography see →sea-floor topography.

bottom water a high-density →water mass that has sunk down to the ocean bottom. The origin of bottom water lies at the polar regions, where water sinks because of convective cooling. Bottom water, however, can be traced all the way to beyond the equator, becoming more and more dispersed into higher levels.

Bouguer anomaly see →gravity anomaly.

Bouma sequence see →turbidite.

boundary current in oceanography the term boundary current is preferentially used for the narrow, strong currents that are found at the western side of the oceanic →gyres. This strengthening of the western side of the gyre, the so-called western intensification, is one of the important issues in the circulation of the oceans. Theoretical explanations are based upon the balance of →vorticity. Examples of boundary currents are the →Gulf Stream along the east coast of N. America and the Kuroshio along the coast of Japan. These currents are associated with a sharp transition between relatively light, warm, and saline water in the east and relatively heavy, cold, and less saline water in the west. Because of dynamic instability these currents form →meanders that may even become detached from the main current and drift around as isolated large →eddies.

boundary layer in a fluid that flows along a boundary a layer develops in which the flow pattern is determined by the transport of momentum between the flow and the boundary. The current in this layer differs in speed and/or direction from that outside this layer, and boundary layers are therefore characterized by current →shear. In these layers →turbulence is generated that is not only effective in the transport of momentum, but also causes transport of other physical and chemical properties between the boundary and the undisturbed fluid.

Boussinesq approximation approximation of the hydrodynamical equations, attributed to Boussinesq (1903), see further →ocean dynamics.

box corer (syn. Kastengreifer), type of →corer that retrieves relatively undisturbed surface sediment samples in a box. The boxes have a variable length and surface. Upon retraction of the box out of the sediment, a spade is rotated over the snout in such a way as to make a tight seal. See also →bottom samplers. (Figure see p. 32)

Bq see →becquerel.

Brachiopoda lamp-shells, phylum of the Lophophorates (see →lophophore). Unsegmented sedentary animals with a two-valved shell, which are usually anchored with a contractile stalk or peduncle. The two shells enclose the body, not laterally like those of →Bivalvia, but dorsally and ventrally. They live mainly fixed to a substrate, sometimes buried in sandy bottoms. They have a large coiled circular lophophore, two spiral projections of the

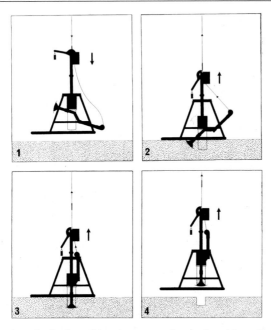

Box corer

anterior body wall bearing a row of tentacles with →cilia for collecting food particles from the water. Brachiopoda were very important in →Paleozoic and →Mesozoic times; at present the number of species is ca. 200.

brackish water seawater diluted by freshwater. See →estuary.

Branchiopoda class of the subphylum →Crustacea (phylum →Arthropoda). Small crustaceans living in freshwater.

Branchiura carp-lice, class of the subphylum →Crustacea (phylum →Arthropoda). Related to the class →Copepoda, small in size, living as ectoparasites attached with the first pair of antennae to the skin of fish (freshwater and marine), where they eat mucus and blood.

breaking of waves see →wave approach.

brines in marine systems occur in surface or subsurface waters containing high amounts of salt due to submarine →hydrothermal vents. Well-known brine spots are found in the Red Sea, where pits occur of more than 1800 m depth filled with brine waters, having a temperature up to 60 °C. These brines are welling up in regions of new crust formation. Brines in surface waters usually occur in shallow semi-enclosed bays in tropical and subtropical regions, due to surplus evaporation.

brittle star see →Ophiuroidea.

brown algae see →Phaeophyceae.

brown clay see →red clay.

Brunhes epoch, chron see →magnetostratigraphy.

Brunt-Väisälä frequency the upper value for the frequency of →internal waves that depends on the vertical stability in the water. This limiting frequency is the result of the balance between inertial and →buoyancy forces, and is also called buoyancy frequency. The Brunt-Väisälä frequency, N, is a measure for the stability. The name comes from the British and Finnish meteorologists, D. Brunt (1927) and V. Väisälä (1925).

Bryozoa (syn. Polyzoa or Ectoprocta, bryozoans) phylum of the Lophophorates (see →lophophore). Though they are not as familiar as other invertebrates, bryozoans are one of the most important phyla with some 4000 extant species. They are small (<1 mm) sessile animals living in colonies that can cover fixed substrates like stones or shells or have a structure reminiscent of moss (hence "moss animals"). They live in freshwater as well as marine, mainly coastal, waters, but some species occur as deep as 8000 m. Colonies consist of individuals specialized in different ways, e.g., the →zooids, which have a protruding lophophore with a circle of tentacles for collecting food particles. The body contains a large →coelom and a U-shaped gut, but other organs are absent. Recent Bryozoa are divided into the orders →Cyclostomata, →Ctenostomata, →Cheilostomata, and the freshwater Phylactolaemata. They have a rich fossil record, of which the orders Cystoporata, Trepostomata and Cryptostomata all became extinct at the end of the →Paleozoic.

Bryozoa. (Barnes 1987)

buffer capacity of seawater seawater is buffered at a relative constant pH range of 7.8 to about 8.3. This buffering is defined by weak-acid forming substances of bicarbonate, carbonate and borate, and to a lesser extent by other compounds. Bicarbonate-carbonate concentrations are in the order of 2.4 mEq dm^{-3} while that for borate is 0 to 5 mEq dm^{-3}, of which only 20 % is ionized at 10 °C. Despite the general buffering effect, some pH variations do occur, influenced by the utilization and production of CO_2 by →photosynthesis and respiration, respectively, of marine flora and marine fauna. See also →alkanity.

buoy in oceanography buoys are used to mark a certain location or body of water, or to carry measuring instruments. They are usually marked red-yellow, differently from navigational buoys. There are also subsurface floats that have been trimmed in such a way that they follow the current at a given level. Some buoys carry a transmitter to send data to a ship or land station, whereas other buoys carry their own data recorder, which should be returned.

buoyancy difference between the gravitational force exerted on a submerged object and the gravitational force acting on the mass of water displaced by this object. This force was already described by Archimedes (287–212 B.C.) and is sometimes called Archimedes force. Positive buoyancy means that the object is lighter, negative that it is heavier than the surrounding water. The concept of buoyancy is also extended to water with a different →density than its surroundings. In that case one may express buoyancy as reduced gravity: $g(\varrho_1-\varrho_2)$ instead of the gravity $g\varrho_1$ (where g is the gravitational acceleration and ϱ_1 and ϱ_2 are densities, ϱ_1 of the water considered, ϱ_2 of the surrounding water). For objects with negative buoyancy one usually speaks of the underwater weight. Pelagic organisms are often capable of regulating their buoyancy. Rapid changes in buoyancy can be performed by the →gas bladder of fish. Gas-filled spaces are also found in various →mollusks (e.g., in the shells of *Nautilus*). An increase in buoyancy can also be obtained when the body accumulates substances with a low specific gravity. For this reason many marine species have a high fat content.

buoyancy force pressure force acting on a water element due to density differences with the surrounding water. Positive buoyancy means that the water is lighter than the surrounding water, negative buoyancy means that it is heavier.

buoyancy frequency see →Brunt-Väisälä frequency.

byssal gland see →byssus.

byssus strong threads secreted by the byssal gland of mussels and other →Bivalvia and used to attach themselves to solid objects.

C

¹⁴C-method, productivity measurement one of the most widely used methods for measuring aquatic plant production in bottles with radioactive carbon added as ^{14}C in bicarbonate. First developed by E. Steeman Nielsen in 1952. After exposure of bottles to natural light (→in situ incubation), or after incubation in experimental incubators for short periods, plankton is filtered from the water, after which the quantity of ^{14}C present in the produced organic substances in the algae is determined, usually in a liquid scintillation counter. Using the known ratio of labeled over total CO_2 present in the bottles at the start of the experiment, the total amount of organic matter produced is calculated from the number of counts. Corrections have to be made for the CO_2 uptake by processes not related to →primary production, which are measured simultaneously in dark incubation bottles. See also →radioisotopes.

Calcarea class of the phylum →Porifera or sponges. Calcareous sponges, generally small in size, living worldwide in the oceans, mostly in coastal areas. The skeleton exists of calcium carbonate spicula; spongine fibers are absent.

calcareous ooze see →ooze.

calcite see →carbonate, mineral.

calcium carbonate see →carbonate, mineral.

calibration in general, adjusting the output of measuring instruments or methods in standarized conditions to the output of a standard method. See also →models and →intercalibration.

caloric equivalent for studies of energy flow, the energy content of (groups of) organisms is expressed in caloric (or calorific) equivalents. The energy content is estimated either by measuring heat at total oxidation (by bomb calorimetry) or by using the known equivalents of the components of the organism in question. Such components are lipids with a caloric value of ± 9.5 kcal g^{-1}, carbohydrates (± 4.2 kcal g^{-1}) and proteins (± 5.7 kcal g^{-1} at total combustion, but only ± 4 kcal g^{-1} at metabolic oxidation within animals). Nowadays, energy content is generally and more correctly expressed in joules (1 cal equals 4.1868 J). Caloric equivalents are high in organisms with a high fat content and low proportions of water and inorganic material. Therefore, caloric values are preferably expressed in joules per gram of either ash-free dry weight or carbon. Roughly, 1 g of carbon yields 12 kcal or 50 kJ.

Cambrian see →geological time scale.

camouflage see →coloration in marine animals.

canyon see →submarine canyon.

capillary wave a surface wave where the →surface tension is the main restoring force. Their wavelength is shorter than 5 cm, because for longer wavelengths the gravity takes over as a restoring force (see →gravity waves).

carapace the shield-like part of the →exoskeleton that covers the thoracal segments in many species of →Crustacea.

carbohydrate(s) a heterogeneous group of compounds with the empirical formula $C_x(H_2O)_y$ (hydrates of carbon). They are prominent constituents of plants, owing to the abundance of cellulose and starch, but they also occur in animals as glycogen, where this serves as the chief source of energy. Carbohydrates, as polyhydroxyaldehydes or ketones and derivatives of them, may be classified into groups according to their complexity. (1) Monosaccharides or simple sugars: single straight chain molecules, not to be hydrolyzed into simpler substances. (2) Oligosaccharides: union of 2 to 12 monosaccharide units. (3) Polysaccharides: union of more than 12 monosaccharide units. (4) Derived carbohydrate oxydation and reduction products, aminosugars and deoxy sugars. The content of carbohydrate in marine organisms varies between ca. 60% on a dry weight basis in brown and red algae, ca. 20% in bacteria, ca. 12% in phytoplankton, and ca. 2% in zooplankton and zoobenthos. Mono- and disaccharides and low molecular polysaccharides excreted by living organisms are readily utilized by microorganisms. Carbohydrates such as cellulose and hemicellulose undergo much slower decomposition.

carbonate compensation depth see →carbonate dissolution.

carbonate dissolution carbonate sediments are rare below a water depth of about 5 km in the Atlantic and about 4.5 km in the Pacific. This distribution depends partly upon the property of $CaCO_3$, that it is more soluble in cold than in warm water, but mainly on the hydrostatic →pressure in the deep ocean. The carbonate equilibrium will shift with increasing pressure towards more →bicarbonate and hydrogen ions, so that the system becomes undersaturated in calcium carbonate derived from skeletons or in carbonate formed →in situ in the upper water layers. Both these effects cause the $CaCO_3$ saturation of seawater to decrease with increasing depth. Below a critical depth, $CaCO_3$ is undersaturated and tends to dissolve. The depth at which calcium carbonate shows a rapid acceleration in dissolution is termed the lysocline. Whether or not calcium carbonate can be preserved at certain depths is a complex function of the rate of deposition versus the rate of dissolution. The depth below which the rate of dissolution exceeds the rate of deposition is termed carbonate compensation depth (CCD). See also →carbonate.

carbonate, mineral mineral compound characterized by a fundamental anionic structure of CO_3^{2-}. The common rock-forming calcium carbonate ($CaCO_3$) exists as the mineral polymorphs calcite and aragonite. Both may form as inorganic precipitates or as biological secretions in the hard parts of organisms. Many multicellular invertebrates, lime-secreting algae and foraminifers produce hard parts of coverings and skeletal carbonate material which eventually become part of ocean sediments or reefs. This process depends only on the chemistry of body fluids or enzymes and may take place even where surrounding waters are barely saturated or undersaturated. Calcite is one of the most common minerals and the principal constituent of limestone; it also occurs in crystalline form in marble, loose and earthy in chalk and stalactitic in cave deposits. The calcite mineral is usually white or colorless, with a perfect rhombohedral cleavage, a vitreous lustre, and a hardness of 3 on the Mohs scale. Aragonite is the high pressure equivalent of calcite; it has a greater density and hardness, and a less distinct cleavage. It is metastable under earth's surface conditions and therefore less common than calcite. Many shells are formed of aragonite, which in time undergoes recrystallization to calcite. Aragonite occurs as a deposit from hot springs and as a major constituent of some marine shells (pteropod), shallow marine muds and the upper parts of coral reefs. The double carbonate dolomite, $CaMg(CO_3)_2$, is mainly a diagenetic mineral formed by the alteration of calcium carbonate through the action of magnesium bearing water. Dolomite is white, colorless or tinged yellow, brown, pink or gray; it has perfect rhombohedral cleavage and a pearly to vitreous luster.

carbonate system →carbon dioxide is dissolved in seawater where it is equilibrated with several other dissolved forms, the bicarbonate ion (HCO_3^-) being very predominant. Equilibration with atmospheric carbon dioxide would also be expected, but in fact the surface waters of the ocean are often either over- or undersaturated relative to the atmosphere. The various forms of carbon dioxide (gaseous CO_2, H_2CO_3, HCO_3^-, CO_3^{2-}) in seawater are related by well-known chemical equilibria and the overall carbonate system is acting as a buffer for the acidity of seawater, maintaining the pH at about the 8.2 level. →Photosynthetic uptake of CO_2 causes the system to shift and the pH to increase. In the surface ocean the concentrations of CO_3^{2-} and Ca^{2+} are sufficiently high for supersaturation of $CaCO_3$ as crystalline calcite or aragonite, allowing organisms to form calcareous shells and platelets. Abiotic precipitation of $CaCO_3$ does not occur. Eventually, the skeletal material settles to the deep sea floor, where sometimes the ambient concentration of CO_3^{2-} is sufficiently low again for $CaCO_3$ to be undersaturated and the skeletons of first aragonite and then calcite dissolve again.

carbon cycle the global transport of carbon atoms (C) between the oceans, the atmosphere, the continents, and several other reservoirs. The overall cycling is very complex, being driven by combinations of physical, biological, chemical, and geological forces. There exist many overlaps with the similar cycling of other chemical elements, such as nitrogen (N), phosphorus (P), sulfur (S), etc., notably with respect to biological processes. Major inorganic forms of carbon are the →carbon dioxide in the atmosphere and dissolved in the oceans, as well as huge sedimentary deposits of biogenic calcium carbonate (shells and their metamorphosed products). By partly absorbing long-wave radiation from earth to space, the atmospheric carbon dioxide has been and still is acting as an insulator, helping planet earth to maintain an overall higher temperature than otherwise would have been possible: the greenhouse earth. Organic forms of carbon constitute all living biomass as well as organic matter in soils and sediments and organic gases like methane in the atmosphere. Through →photosynthesis, plants (on land) as well as algae (in the sea) are converting inorganic carbon dioxide into an immense variety and complexity of organic compounds. Most of this is being recycled by animal respiration or bacterial breakdown, but a small portion escapes and throughout geological times this has led to the formation of reservoirs

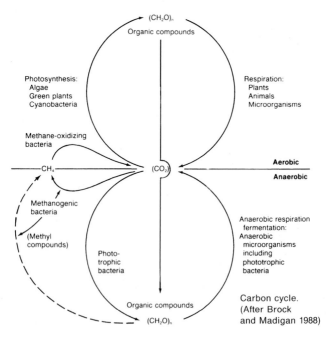

Carbon cycle.
(After Brock
and Madigan 1988)

of organic →carbon, as coal, petroleum, and natural (methane) gas. The increased rate of combustion of such fossil fuels by mankind has led to the rapid increase of atmospheric →carbon dioxide, at the risk of rapidly enhancing the insulating or "greenhouse" effect, possibly leading to global warming. Sources are less well defined for atmospheric levels of methane, which are also rising rapidly and similarly affecting the heat balance of the earth. Throughout geological times relatively slow shifts in the carbon cycle are known to have been critical for corresponding climatic changes.

carbon dioxide (CO_2) gas resulting from oxidation (e.g., combustion) of elemental carbon (C) or reduced (i.e., organic) C; the only stable form of C when following the laws of chemical thermodynamics, which, however, only apply to equilibria in closed systems. Earth is an open system and the continuous flow of solar energy has given rise through →photosynthetic conversion of CO_2 to the existence of all biota: the living planet. Carbon dioxide is dissolved in seawater where it is equilibrated with several other dissolved forms, the bicarbonate ion (HCO_3^-) being very predominant. The total amount of various forms of carbon dioxide (gaseous CO_2, H_2CO_3, HCO_3^-, CO_3^{2-}) in ocean waters is 40 10^{18} g C, exchanging at an annual rate of about 100 10^{15} g C with the 700 10^{15} in the atmosphere. The latter corresponds to about 0.0353 % of the atmosphere consisting of carbon dioxide. This atmospheric concentration of 353 ml m^{-3} in 1990 is higher than ever before in the past 160 000 years (as estimated from air bubbles trapped in the Antarctic ice sheet) and increasing at an annual rate of about 2 ml m^{-3} as a result of fossil fuel burning and rapid deforestation, possibly giving rise to global warming. Currently, mankind introduces CO_2 at about 5.1 10^{15} gram C per year from fossil fuel burning, presumably another amount of CO_2, about 2 10^{15} g C per year, by deforestation. Of this total input of about 7 10^{15} g C per year only about 2 10^{15} g C is retained in the atmosphere (causing the increasing CO_2 concentration), the remaining about 5 10^{15} g C presumably annually being taken up by the oceans, slowly adding to the already immense reservoir of CO_2 in seawater.

Carboniferous see →geological time scale.

carbon, organic virtually all chemical compounds with the reduced C atom (valency −4) as major constituent. Molecules almost always also contain H, O, as well as sometimes N, P, S, and also trace amounts of other elements. On earth virtually all these have been formed on the basis of →photosynthesis, that is the conversion of CO_2 by plants (on land) as well as algae (in the sea) with the required energy provided by sunlight. Exceptions are submarine hydrothermal ecosystems which are supported by geothermal energy. All living biota, whether plants, animals, or bacteria consist of organic compounds (e.g., carbohydrates, proteins, lipids) and are the natural precursors of "dead" organic matter. The latter may be dissolved organic carbon (DOC) in seawater (including large complex humic and

fulvic acids), organic matter stored in marine sediments including petroleum, natural gas (methane), as well as organic matter in soils on land,
which may finally become preserved as peat or coal. Living organic matter
consists predominantly of chains of C atoms (aliphatic molecules), whereas
dead natural organic matter may contain relatively more C-rings (aromatic
molecules). For centuries it was believed that organic carbon compounds
can only be formed by organisms, yet from 1783 (potassium cyanide by
Scheele) onwards (e.g., ureum by Wohler in 1828, later more complex
compounds by Berthelot), organic synthesis has evolved rapidly in the
laboratory as well as in manufacturing. This has led to the introduction of
many new organic compounds in the marine environment, including the
generally toxic classes of halogenated (chlorinated, brominated) organics.
See also →carbon cycle.

carbon, radioactive due to cosmic radiation from sun activities, a number of
natural →radionuclides are formed in the atmosphere. Radioactive carbon
such as ^{14}C is formed due to neutron capture by nitrogen, whereby stable
^{14}N is transformed into ^{14}C. The approximate rate is $1.8-2.5$ atoms
$cm^{-2} s^{-1}$. ^{14}C has a half-life of 5730 ± 40 years, and is therefore used as a
tool to determine the age of organic material formed up until 40 000 years
ago, on the basis of its $^{14}C/^{12}C$ ratio. Because, however, many radiocarbon
dates have been published which are based on the old half-life of 5568 years,
this value is still used conventionally. In oceanography, ^{14}C has been widely
applied to determine the age of water masses by the isotope ratio of
dissolved total CO_2, and is subsequently an estimate of deep-sea circulation
patterns of major ocean basins. ^{14}C age determinations have as interfer-
ences the twofold addition of ^{14}C derived from bomb-test nuclear
explosions from 1945 until 1963, and CO_2 containing no ^{14}C from fossil fuel
combustion. See also →radio isotopes.

carnivores generally all animals that eat other animals; in food-chain studies
secondary consumers (those feeding on primary consumers i.e., secondary
producers or →herbivores) and all animals in the higher trophic levels.

carotene see →carotenoids.

carotenoids yellow, orange or red fat-soluble pigments. Almost all these
molecules are based on the formula $C_{40}H_{56}$. The carotenoids are usually
classified into carotenes, which are hydrocarbons, and xanthophylls, which
are oxygenetic derivatives of carotene. Carotenoids absorb light in the
region of 400 to 550 nm. A very common xanthophyll is fucoxanthin, which
is found in diatoms in such quantities that these organisms are not green but
brown.

carrying capacity the maximum population size (→density of an organism)
possible in an →ecosystem, beyond which the density cannot increase
because of environmental resistance.

cast in oceanography a cast is a series of water samplers lowered on one line, to obtain (roughly) simultaneously water samples at different depths.

catadromous downward running, term used to indicate fish species of freshwater that make spawning migrations to the sea, such as, e.g., the anguilliformes or eels.

cellulases group of enzymes, each catalyzing a different step in the depolymerization of cellulose into the disaccharide cellobiose or into glucose. The extra-cellular enzymes are widespread in cellulose-decomposing microorganisms.

cellulose most widely distributed skeletal polysaccharide (in plants), usually associated with other compounds such as lignin and hemicellulose. The molecule consists of 4000 to 7000 cellobiose-units (a disaccharide), has a molecular weight of 1 to 2 10^6, and is insoluble in water. Depolymerization is mainly performed by microorganisms.

Cenozoic see →geological time scale.

centric diatoms see →Baccilariophyceae.

centrifugal force a fictitious force that has to be introduced in the balance of forces if this is related to a rotating coordinate system. In mechanics, a rotating body has to be continuously accelerated toward the centre of rotation by a so-called centripetal force that counteracts the effect of inertia, in order to keep this body in its circular path of motion. If, however, the balance of forces is considered relative to a rotating coordinate system (as we normally do on the rotating earth) the inertia appears as a radially directed force (the centrifugal force) that must be balanced by the centripetal force exerted. On the earth, the centrifugal force is combined with the gravitational force in the values given for the acceleration of gravity.

Cephalochordata subphylum of the phylum →Chordata. They are chordates with an elongated, fish-like shape, but unlike the vertebrates they have no skull, brain, or any cartilaginous or bony skeleton. Characteristic are the typical chordate tube-like dorsal nerve-cord with double nerve roots, the muscle blocks (myotomes), and the gill slits. The sexes are separated and there is no asexual budding. Example: the lancet fish *Amphioxus*.

Cephalopoda (cephalopods) class of the phylum →Mollusca, including *Nautilus*, cuttle-fish, squid, and octopods. Generally free-swimming raptorial mollusks, among which are the largest of all invertebrates (the giant squid, with a length of up to 16 m). Typical is the circle of arms or "tentacles" with rows of suckers, on the head, which developed from the mollusk foot (see →Mollusca). They swim by jet propulsion: water is pressed out of the mantle cavity through a funnel. The shell, external in Ammonoidea (extinct) and Nautiloidea (almost extinct), is in other

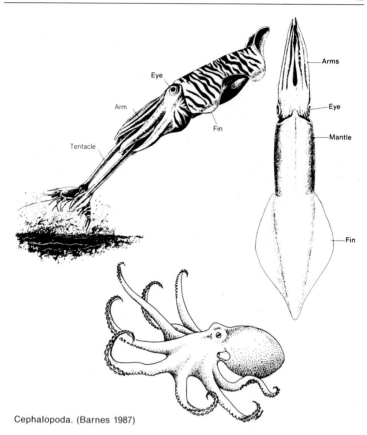

Cephalopoda. (Barnes 1987)

cephalopods reduced and internal, or absent. Cephalopoda have a mouth with a horny bill, and highly developed eyes. In the skin are →chromato-phores which enable the animals to change their color quickly (alarm response) and also light organs. Examples are *Loligo, Sepia, Octopus* and *Nautilus*. Although there are at present still several hundreds of species, their peak development was during the →Mesozoicum. There are 7500 fossil species (cephalopod shells), e.g., the well-known →ammonites.

Ceriantharia order of the subclass Zoantharia, class →Anthozoa (phylum →Cnidaria). Anemone-like anthozoans with a large number of tentacles in two series. The body is elongated and adapted for living in sand or mud in secreted tubes.

Cestoda class of the phylum →Platyhelminthes or flatworms. Parasitic flatworms, in which the gut is absent. Examples are the tape worms (*Taenia*).

Cetacea (cetaceans), see →marine mammals.

chaeta (pl. chaetae), chitinous stiff hair or bristle, present in, e.g., →Chaetognatha, →Polychaeta and →Oligochaeta.

Chaetognatha (chaetognaths, arrow worms) phylum. Elongated (to a few centimeters), transparent, unsegmented →coelomate animals with a marked head, trunk, and tail. The head bears eyes, chitinous spines, and jaws. The trunk and tail have lateral and caudal fins. The alimentary canal is straight and simple, there is no special excretory, respiratory, or vascular system. Arrow worms are cross-fertilizing →hermaphrodites without larval stage. Some species are characteristic for certain water masses and are as such indicators of plankton communities. Example: *Sagitta*.

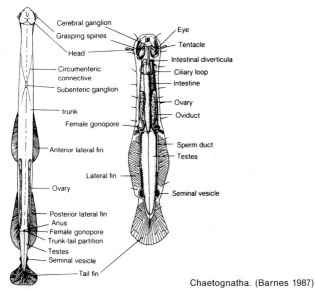

Chaetognatha. (Barnes 1987)

Challenger Deep Sea Expedition the first global survey of the deep ocean, directed by Charles Wyville Thomson. The 2300 tons corvette Challenger with auxiliary steam power covered nearly 70000 miles in the 1872–1876 period. Over 500 deep soundings, 133 dredgings and various other operations were executed at 362 oceanographic stations. As a result, 715 new biological genera and about 4500 new species were described. Utilizing steel hydrographic wire, a record deep-sea sounding of 8180 m in the →Mariana Trench was recorded. W. Dittmar analyzed very accurately 77 seawater samples collected by Buchanan from which the concept of the "constant composition of seawater" was derived, which is the concept underlying salinity being a cornerstone measure for marine sciences. The initial results of the expedition were published in *The Voyage of H.M.S. Challenger* (1884), edited by Sir John Murray, who in this and later publications laid the foundation for the field of marine geology. The expedition marks the beginning of modern oceanography, and for the 70 years following the expedition its reports were the major source of information on the deep ocean.

channel (1) relatively narrow sea or stretch of water, wider and larger than a strait, connecting two larger bodies of water. Also used for the English Channel between England and France; (2) deeper part of bottom below moving water (in a bay, estuary or strait) through which the main current flows.

Charophyceae class of the green plants or →Chlorophyta; candleworths and related species. →Algae that approach the higher plants in level of organization. Charophyceae contain the pigment →chlorophyll-b and lutein. They store carbohydrate as starch inside the plastids. The flagellar apparatus of their →zooids, including the parts inside the protoplast, resembles in fine structure the corresponding organelles in mosses and ferns. During cell division, the primary cell wall develops in the same way as in higher plants, different from all other organisms. Though this class chiefly comprises freshwater plants, the genus *Chara* also lives in brackish marine environments, for instance in sheltered shallow bays of the Baltic.

chart datum a reference level for chart depths. The chart datum needs not to be →mean sea level or horizontal plane, but is usually connected with the local tide, e.g., the mean lower low water level at spring tides.

Cheilostomata order of the class Gymnolaemata (phylum →Bryozoa). Colonies composed of →zooids are generally flattened and have calcified body walls. The aperture of the zooids is closed by an →operculum. In many species modified zooids (→vibracula, avicularia) occur. Example: *Bugula.*

chelation a term generally used for all kinds of reversible reactions between dissolved metal and organic matter, the organic matter being either in dissolved or solid form. In fact, chelation relates to stereometric claw-like binding of a metal by an organic molecule. This type of reaction is known in ion-exchange resins, but also in dissolved organics such as EDTA (Ethylene Diamine Tetra Acetic acid-disodium salt). Seawater contains a number of dissolved organic substances which are able to complex (chelate) metals, such as copper. This complexing has to compete with all major and minor cations present. It is essential to understand the complete system of chelation, using, if available, the →stability constants for each organic-metal combination. See also →complexation.

cheliped see →Crustacea.

chemical oxygen demand (COD) is an index of the total organic carbon in a water sample. The COD is determined by measuring the amount of oxidizing reagent consumed during oxidation of organic matter with dichromate or permanganate. See also →biological oxygen demand (BOD).

chemical weathering the breakdown and dissolution of continental rocks, both igneous and sedimentary, under the influence of rain and soil water. The dissolved chemical elements become constituents of water in streams and rivers. The chemical composition of river water varies considerably between various rivers, reflecting the different rock composition of continental drainage basins. Eventually the river water enters the ocean, contributing the major constituents (dissolved salts) in seawater.

chemoautotrophic bacteria see →metabolic diversity.

chemolithotroph see →metabolic diversity.

chemoorganotroph see →metabolic diversity.

chemostat a continuous culture device for microorganisms controlled by the limiting concentration of a nutrient or other substance.

chemostratigraphy stratigraphy based on the chemical composition of rocks, such as the content of →rare earth elements and →organic carbon, →alkalinity, etc. See also →stratigraphy.

chemotaxis the movement of an organism towards or away from a chemical substance. Positive chemotaxis is towards an attractant; chemicals that induce negative chemotaxis are repellents.

chert lithified siliceous →ooze.

chitin in the marine environment the commonest bio-polymer, probably second only to cellulose as the most abundant biogenic substance on earth. Chitin belongs to the group of mucopolysaccharides, i.e., polysaccharides built up by amino-sugars (sugars in which a hydroxyl group is replaced by an amino group), in the case of chitin: N-acetyl glucosamines. It forms the hard shell of crustaceans and insects, and is in many respects analogous to cellulose.

chlorinity in 1902 the Swedish chemist S.P.L. Sørensen gave the definition: "Chlorinity stands for the weight of the collected halogens present in one kilogram of seawater". Chlorinity is chemically determined by argentometric titration. In 1940, Knudsen and Jacobsen redefined: "The number giving the chlorinity in permille of a seawater sample is by definition identical with the number of gramatoms silver just necessary to precipitate the halogens in 0.3285234 kilogram of the seawater sample". Until 1966 chlorinity was used to calculate →salinity. The basis of the salinity estimation from the determination of a single major element is the long-known concept of constant relative composition of seawater.

chlorite a group of platy, greenish clay minerals consisting of a 2:1 layer (see →clay minerals), whose negative charge is balanced by an extra interlayer consisting of a positively charged octahedrally coordinated hydroxide sheet. The structure unit thus contains two octahedral sheets and two tetrahedral sheets. The general formula is: $(Mg, Fe^{2+}, Fe^{3+})_6AlSi_3O_{10}(OH)_8$. Chlorites are distributed in low-grade metamorphic rocks or as alteration products of ferromagnesian minerals. See also →clay mineral.

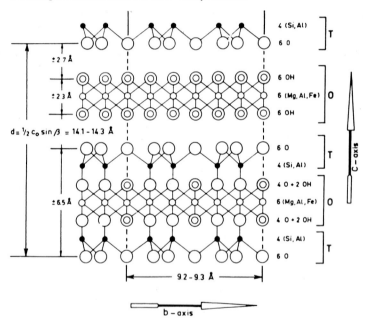

Chlorite. (After Beutelspacher and van der Marel 1968)

Chlorobacteriaceae see →Chlorobiaceae.

Chlorobiaceae (or Chlorobacteriaceae) →phototrophic green bacteria.

Chloromonadida order of the class Phytomastigina (subphylum →Mastigophora, phylum Sarcomastigophora, see →Protozoa), which have numerous green →chromatophores or are colorless. They have reserves of oil and a gullet or complex contractile vacuole. No transverse groove. They possess a delicate pellicle, or are ameboid.

Chlorophyceae class of the green plants or →Chlorophyta, which comprises the green algae ranging from →picoplankton to seaweeds of more than 1 m long. They contain the pigments →chlorophyll-b and lutein, and store carbohydrate as starch inside their plastids. In *Caulerpa*, cell walls are only formed to separate reproductive parts, and numerous nuclei share the vegetative compartment. *Ulva* has one nucleus per cell. Most green seaweeds are intermediate in this respect. *Pyramimonas* and *Tetrahele* are important in estuarine and coastal phytoplankton. Cells with walls of sporopollenin are part of some chlorophycean life histories. Some genera are calcium carbonate-encrusted. These characteristics are responsible for a fossil record going back to the →Paleozoic.

chlorophyll (leaf-green) a complex of pigments causing the green color of plants and giving them the ability to assimilate CO_2 from the air (or water). Chlorophyll is situated in →chloroplasts in the protoplasm. The pigment has its strongest light absorption in the red and blue-violet (hence its color). In the presence of the enzyme system related with chlorophyll, part of the absorbed light energy is stored as potential energy, i.e., the first step in →photosynthesis of carbohydrates from CO_2. See also →pigments, plants.

Chlorophyll

Chlorophyta green plants, a division of the →eukaryotes comprising all higher plants, the mosses, and a number of algal classes being the →Chlorophyceae, the →Prasinophyceaea and the →Charophyceae. See also →algae.

chloroplast see →chromatophore.

choanocyte (collar-cell) peculiar kind of cell with a →flagellum surrounded by a thin protoplasmatic sheath or collar, found only in sponges and a small group of Flagellata (Choanoflagellata).

Chondrichthyes cartilaginous fish, see →Pisces.

Chordata (chordates) phylum of the animal kingdom containing the subphyla Urochordata (tunicates), Cephalochordata (lancelets) and Vertebrata. These share three important characteristics: (a) there is, at least during the embryonic stage, a →notochord running from head to tail in the back of the animal; (b) there are, during some developmental stage, gill slits (lateral openings from the →pharynx to the external) and (c) there is a hollow dorsal nerve cord.

chromatophore (1) cell with pigment granules in its cytoplasm. Rapidly changing disposition (concentration or dispersion) of the pigment within these cell alters the color of the animal; (2) chromoplast, a plastid (protoplasmic inclusion) carrying one or several →pigments, either →carotenoids or →chlorophyll. In the latter case the plastid is called chloroplast.

chromoplast see →chromatophore.

chronostratigraphy (syn. time stratigraphy) branch of stratigraphy that deals with the age of strata and their time relations. See →stratigraphic hierarchy and →stratigraphy.

Chrysomonadida see →Chrysophyceae.

Chrysophyceae (syn. Chrysophyta) class of the division →Heterokontophyta, golden-brown algae including the →silicoflagellates), class of mostly

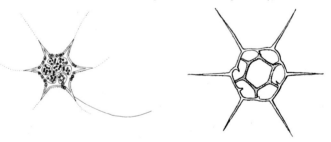

Chrysophyceae. (Riedl 1963; van den Hoek 1984)

microscopic →algae which in many cases produce urn-shaped, siliceous resting spores. Chrysophyceae contain →chlorophyll-c and →fucoxanthin. They accumulate fluid chrysolaminarin in vacuoles as a carbohydrate reserve. Few of them are marine. The colorless genus *Monas*, with reduced plastid, feeds on →picoplanktonic algae. *Dictyota* and *Distephanus*, with internal skeletons of silica, are widespread in marine phytoplankton. Siliceous remains of Chrysophyceae are abundant in some marine sediments of →Tertiairy origin. Chrysophyceae are in principle protists (see →Protozoa), which can also be listed as the order Chrysomonadida in the subphylum →Mastigophora.

Ciliata (ciliates) see →Ciliophora.

Ciliophora (syn. Ciliata) phylum of unicellular heterotrophic organisms belonging to the kingdom Protista (see →Protozoa). The main characteristic of this homogeneous group is the possession of →cilia with coordinated movement that propagate the cell through the water, and transport food particles to the mouth or cytostoma. The body shape is normally constant and asymmetrical, and in some groups of ciliates housed inside a basket-like house or lorica, excreted, and often composed of foreign materials like →coccoliths. Ciliophora are widely distributed in marine and freshwater. In the sea they are important in shallow microbenthic foodwebs, and also in the microzooplankton, where the suborder →Tintinnina is of interest. (Figure see p. 50)

cilium (pl. cilia) fine hair-like vibratile protension on the surface of a cell. A cilium has the same structure as a →flagellum, but is much shorter. Each cell carries many cilia; a sheet of such cells makes a ciliated epithelium. All cilia on a cell as well as in an epithelium move in a coordinated rhythm. During the effective phase of a beat the cilium moves stiffly through the medium (water, mucus), exerting a force on it, while at the recovery stroke the cilium is bended, offering less resistance. By this action the organism propels itself through the water (e.g., →Ciliophora) or it transports water into or in its body, as many →filter feeders (e.g., →Bivalvia, →Tunicata) do.

circadian rhythm see →biological clocks.

circulation in the ocean is determined by the wind field over the ocean and by →thermohaline processes. The large-scale circulation which underlies the time variations in the circulation, is characterized by large →gyres in the upper layers under the influence of the mean wind stress, and by a deep water circulation (roughly below 1000 m depth), in which under the influence of convection the water sinks in more or less localized source areas of high-density →water masses. This deep water subsequently spreads over the different ocean basins, until it gradually reaches the upper layers again. The patterns of the ocean circulation are studied in →ocean dynamics, where the contributions of →geostrophic balance, →drift

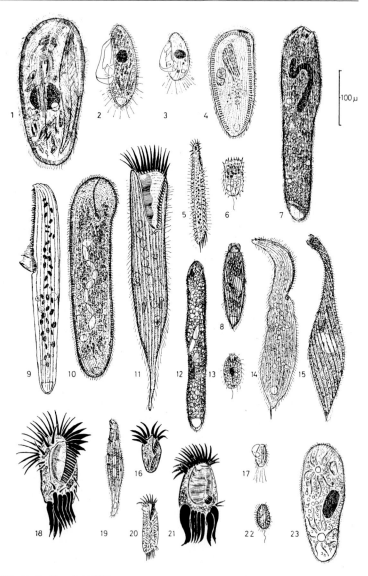

100 μ

Ciliophora. (Fenchel 1969)

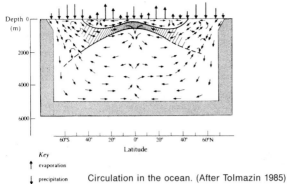

Depth 0 (m)

2000

4000

6000

60°S   40°   20°   0°   20°   40°   60°N

Latitude

Key

↑ evaporation

↓ precipitation   Circulation in the ocean. (After Tolmazin 1985)

currents and of →vorticity balance on the rotating earth play an important
role. Large sets of observations of temperature and salinity, collected by
oceanographic expeditions, surface current data, formerly mainly from
→dead reckoning values from merchant ships, but now more and more
from drifting →buoys, provide the observational evidence. The large-scale
ocean circulation is important in the redistribution of heat and salinity over
the globe.

Cirripedia barnacles, and stalked or goose barnacles, class of the subphylum
→Crustacea (phylum →Arthropoda). They are the only crustaceans that
live on fixed substrates, and also as ectoparasites on animals, from crabs
and mollusks to whales. Fertilization is effectuated by a long tube (penis)
conducting the sperma from male to female. Barnacle larvae (cypris larvae)
live in the →pelagic, where they sometimes constitute an important part of
the zooplankton (see →meroplankton). After settlement on suitable
substrate, to which the animal attaches with cement secreted by cement
glands, the →carapace of the larvae develops into the enveloping carapace
or mantle of the barnacle, forming calcareous plates. Inside this house the

Cirripedia. (Barnes 1987)

crustacean lives upside down, collecting food by sweeping the limbs, extended with hairs (cirri, hence their name), through the water. If disturbed, or in tidal areas at emergence, the animal retracts and closes the carapace.

Cladocera (cladocerans) water fleas, suborder of the order Diplostraca, class →Branchiopoda (subphylum →Crustacea, phylum →Arthropoda). Group of planktonic crustaceans found mainly in freshwater, e.g., *Daphnia.* Marine species occur, e.g., *Podon,* living in the zooplankton.

clams see →Bivalvia.

Clark-Bumpus zooplankton sampler see →zooplankton samplers.

clay (1) detrital particle of any composition smaller than 2 µm, which is the upper limit of size of a particle that can show colloidal properties; (2) sediment composed primarily of clay-size particles and characterized by high plasticity and by a considerable content of clay minerals. See also →grain size and →clay mineral.

clay mineral a term for the constituents of a clay which give it its plastic properties. Clay minerals are concentrated in a size less than about 2 µm. They are layer-lattice minerals occurring as minute, platy, sometimes curly or fibrous particles. The 1:1 layer type (one tetrahedral layer and one octahedral layer) or 2:1 layer type (two tetrahedral layers and one octahedral layer) clay minerals have silicon in tetrahedral and aluminum in octahedral coordination. Silicon and aluminum can be replaced by other ions of a different valency providing a charge to the clay mineral layers. The excess charge within the layer may cause adsorption of exchangeable cations (Ca, Na, Mg, K, etc.). Clay minerals are formed chiefly by weathering and hydrothermal action. The most common mineral groups are →kaolinite, →illite, →smectite, and →chlorite.

clearance rate see →filter feeders.

climate the totality of the weather conditions at a certain location over a certain period (conventional: 30 years). The study of the climate involves the description and classification of the weather as well as the investigation of the physical basis for the climate and its differences and variations. One may describe the climate according to certain larger climatic zones, whereas the effect of local conditions is reflected in the meso- or microclimate. The marine climate is strongly influenced by the →air-sea exchange and is more homogeneous than the climate on land. In the tropics (about 10°N-10°S) light winds and high temperature and humidity prevail (Doldrums). To the N and S the following zones are found: the trade winds, up to 30° with persistent NE- or SE-ly winds and with a chance of →tropical cyclones, the arid "horse latitudes" with light winds up to 35°, the →westerlies up to 60° with from west to east traveling disturbances and large weather variations, and the →easterlies from 60° polewards, where the predominant wind direction changes to east. Geographic conditions modify this pattern, especially in the →monsoon region.

climatic variation the climate is not stable. Studies of historical weather records and geological evidence, such as, e.g., →isotope stratigraphy, show that there have been significant changes in the course of time, sometimes leading to a →glacial epoch. External causes, like →orbital variations cannot always be indicated. For instance, the →El Niño effect in the tropical Pacific appears to be a natural oscillation involving the interaction between the atmosphere and the ocean. During the last decades there has been increasing concern about the effects of deforestation, overgrazing, desertification, rise of →carbon dioxide concentration of the atmosphere, etc. on the climate on a global scale. Carbon dioxide in the atmosphere, together with water vapor and other gases, absorbs part of the outgoing long-wave infrared radiation. An increase such as has been observed over a long period of years will lead to an absorption of emitted and backscattered infrared radiation (→greenhouse effect) and may result in higher temperatures. In evaluating this effect quantitatively, the oceans play an important role as a sink of carbon dioxide, while equally the resulting changes in the →circulation of the ocean may be of influence.

climax community see →community.

clone an assemblage of genetically identical organisms derived by asexual or vegetative multiplication from a single sexually derived individual. In molecular biology a number of copies of a DNA fragment to be replicated by a phage or plasmid.

closure problem the equations of →hydrodynamics are nonlinear and thus couple motions at different scales. One of the objectives of hydrodynamics is to simplify these equations when focusing on a feature occurring at a particular scale. The impact of smaller-scale motions on the relevant scale is solved by →parametrization. In particular this applies to the parametrization of the small-scale nonlinear terms in the momentum equation, the →Reynolds stresses, the description of which in terms of large-scale parameters is known as the closure problem.

Clypeastroidea order of the class →Echinoidea (phylum →Echinodermata) with strongly flattened test. Examples: true sand dollars, sea biscuit.

Cnidaria (syn. Coelenterata, cnidarians) comprising the polyps, jellyfishes, sea-anemones, and →corals. Solitary or colonial, sedentary or free-living animals which are fundamentally radially symmetrical. The body consists of three layers: →ectoderm, →endoderm, and between these the →mesogloea. No →coelom. The Cnidaria have alternatingly a sessile vegetative (→polyp), and a swimming generative (→medusa) generation, though it depends on the species which of the generations dominates the life cycle. The polyps consist of a basal disk for attachment, a hollow body stalk with the

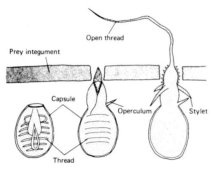

Cnidaria. (From Barnes 1987 after Holstein and Tardent 1984)

gastro-vascular cavity, and a mouth, around which a circle of tentacles. The polyps multiply by asexual budding, which is the development of evaginations on the body stalk as extensions of the gastro-vascular cavity, after which mouth and tentacles form, and hence the new polyp. Many cnidarians (e.g., the corals) form colonies with differently specialized polyps. In other cases, the polyp buds loosen (strobilation) and start a free-swimming life phase as medusa or jellyfish. The body formed is then an upside-down polyp, with the stalk expanded to a large disk or umbrella, and the mouth surrounded with sometimes very long tentacles hanging down. In the ectoderm, especially on the tentacles, are specialized cells, cnidoblasts (typical of the Cnidaria), containing nematocysts which are small organelles that can fire a spirally wound stinging thread with poison. When cnidoblasts are touched, the threads are fired to penetrate and lame the prey, which is transported to the mouth by the tentacles, and digested in the gastrovascular cavity. Cnidaria are found in all seas and oceans with corals in the tropics and some jellyfish like *Aurelia* all over the northern hemisphere.

cnidoblast thread cell, see →Cnidaria.

cnoidal wave a wave that cannot be described by a sinusoidal function, but that is characterized by sharper crests and flatter troughs. The shape of such a wave results from nonlinear dynamics. The name "cnoidal" is derived from the mathematical function "cn" that described this particular shape.

　Cnoidal wave

coast strip of land of indefinite width that extends along the shoreline; in a more restricted sense, a strip of land from the low tide line landward up to the first major change in land features. Coasts include rocks and cliffs, islands, beaches, dunes, reefs, coastal swamps, mangrove, →salt marshes, river mouths, →estuaries and large-scale features such as →deltas, plains, headlands, and bays. Various classifications of coasts have been made, worldwide as well as regionally or locally, based on (a) the degree of erosion (retreat) and deposition (accretion); (b) tectonic uplift (emergence) or subsidence (submergence); (c) the degree of regression or transgression; (d) wave dynamics (protected, sheltered, exposed, wind waves, →swell, monsoon influence; (e) tectonic structures (plate collision, trailing plate edge, marginal sea); (f) the genetic history (primary: drowned coasts, deposition from land, volcanic, fault or fold-shaped; and secondary: wave-eroded marine depositional, built up by organisms) and (g) equilibrium or nonequilibrium. The various classifications summarize the knowledge of coasts in a systematic way. The littoral zone or coastal zone is the zone along the coast, between the extremes of high and low tide (intertidal zone). →Fjords are regarded as coastal features even where they extend far inland. The coast in the wider sense is also called the coastal zone, whereas the littoral zone is more restricted to the shoreline. Usually features of coastal deposition (beaches, →spits, →barrier islands) alternate with erosional features. An important process along many coasts is the along-shore displacement of sediment by longshore current or, in the breaker zone, longshore drift. →Regression or →transgression of the coast is often indicated by submarine or →marine terraces.

coastal upwelling when a →drift current transports water away from the coast, the deficit that occurs will be supplied from deeper layers under the influence of the pressure gradient that develops. It even is not necessary that the wind blows →offshore: the →Ekman transport being at right angles to the wind can also result in offshore transport when the wind blows more or less parallel to the coast, if this coast is to the left of the wind in the northern hemisphere or to the right of the wind in the southern hemisphere. The process of upward moving water is called upwelling. As the water at greater depth is colder, upwelling areas are characterized by relatively cold surface

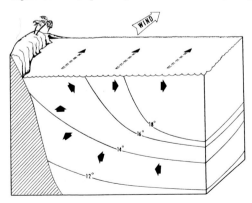

Coastal upwelling.
(After Fairbridge 1966)

water. Upwelling also carries fresh nutrients into the →euphotic layer and therefore promotes the productivity in these regions. Typical upwelling regions are the coasts of Morocco and Namibia in the Atlantic Ocean, and the coasts of Chili and California in the Pacific.

coastline, legal aspects for marking the zones landward and seaward, on which states have certain sovereign rights, the coastline is taken as a base for defining the baseline separating these zones. The normal baseline, according to →UNCLOS III, will be the low-water line (which may be the average low-water line or the average spring low-water line) along the coast, as marked on large-scale charts officially recognized by the coastal state. When the coastline is deeply indented and cut into, when there is a fringe of islands along the coast in its immediate vicinity, or the coastline is highly unstable due to the presence of a delta and other natural conditions, straight baselines may be drawn, joining appropriate points and following the general direction of the coast. Straight baselines may also be drawn across the mouth of rivers, directly flowing into the sea and bays of which the distance between the low-water marks of the natural entrance points does not exceed 24 →nautical miles.

Coccolithophorida (coccolithophores) (see →Haptophyceae), →nanoplanktonic flagellate algae with a layer of minute calcareous plates (→coccoliths). It was only after the invention of the →scanning electronic microscope that study of the form and structure of these plates and proper identification became possible. They may form algal blooms coloring the seawater whitish. Coccoliths are important →microfossils, used extensively in →stratigraphic correlation. Coccoliths may form the bulk of →Cretaceous rocks.

coccoliths microscopic calcareous plates secreted by and covering →Coccolithophorida, averaging about 3 μm in diameter. See also →nanofossil.

coccus spherical bacterium.

COD see →chemical oxygen demand.

Coelenterata (coelenterates) see →Cnidaria.

coelenteron enteron or cavity in →Coelenterata, having a single opening for ingestion and egestion. Often complicated by the presence of partitions separated by so-called →mesenteries, or by extensions (→diverticula), or canals transporting food through the coelenterate body.

coelom (syn. perivisceral cavity) main body cavity in which the gut is suspended. Situated in the →mesoderm.

cold-blooded animal see →poikilotherm.

coliforms, (colibacteria) as indicator of pollution bacteria from the digestive tract of man and warm-blooded animals. Their presence in water indicates that the water is contaminated with fecal material. Hygienic evaluation of drinking and swimming water is made by the determination of the density of these fecal indicators; the coliforms themselves are harmless, but they may indicate the presence of harmful bacteria like cholera bacteria from the same source.

colonization process of establishing populations of one or more species in an area or environment where the species involved were not present before.

color of seawater visually the color of the sea may be affected by the reflection at the sea surface of the light from the sky or from clouds. Besides this there is the contribution by the upwelling light coming through the sea surface. This is the result of a combination of absorption and scattering of light which differ at different wavelengths. In very clear water the scattering is relatively high for blue light, whereas the absorption is relatively high for red light. This gives a blue color. Dissolved substances and suspended particles may change this relation and give a different color.

coloration in (marine) animals for animals colors generally function as →inter- or →intraspecific signals. Interspecific signals are functional in recognizing conspecific animals, animals of different sexes, and play a role in courtship behavior. Interspecific signals can be divided into different categories. (1) Cryptic coloration (camouflage): animals resemble their natural background, or even highly resemble the color and texture of their natural prey, on which they live. (2) Aposematic (warning) coloration: species are highly colored and conspicuous in their natural environment, indicating their unpalatability. (3) →Mimicry: one or more palatable species mimic an aposematic model (Batesian mimicry), or several aposematic species share the same color pattern (Müllerian mimicry). (4) Flight or flash coloration: animals increase their chances of escape by means of flash colors.

comb-jellies see →Ctenophora

commensalism relationship between organisms in which one takes a part or the rests of the food collected by the other organism. See also →parasites.

community assemblage of →populations living in a given area with mutual interactions. Such communities are classified according to dominant species, →indicator species, physical habitat, or type of →community metabolism (e.g., based on rate and efficiency of production). Communities are often only to a certain degree stable. They may change drastically after certain catastrophes (e.g., oil spill) or they may gradually change in a sequence of successions with different dominant species. The succession ends in a climax community, which often has the highest species diversity.

community metabolism total respiration in a community, but also used in a more general sense. See also →community.

compartment models →ecological models that have been subdivided into a number of compartments. These compartments may be constituents or state variables. The term is also used to denote models that are subdivided into spatial compartments.

compensation depth (1) depth at which the light intensity (compensation light intensity) is just enough to balance →photosynthesis (oxygen production) and →respiration (oxygen consumption) at a given moment; varies from species to species; (2) depth below which the rate of dissolution of →carbonate exceeds the rate of deposition, see further →carbonate dissolution.

compensation point depth at which oxygen production (→photosynthesis) is insufficient to meet respiration requirements for 24 h. Sometimes, however, considered equivalent to →compensation depth (light).

competition, biological the struggle for food, space or other ecological requirements between two organisms that have similar ecological needs while neither can realize its requirements fully, and/or thereby interfere mutually with breeding and survival. Such competition between species is called interspecific competition, within species, intraspecific competition. Competition may be direct, as in the case of interspecific territoriality, and is termed interference competition. More indirect competition, such as that arising through the joint use of limited resources, is termed exploitation competition. Because it is always advantageous for either party in a competitive interaction to avoid the other whenever possible, competition presumably promotes the use of different resources and hence generates ecological diversity. Competition plays a major role in →niche theory, as resource utilization spectra (which are determined by competition) largely identify an ecological niche.

complexation a term describing all kinds of organic or inorganic complex formations possible. Such formations may be complexes of great stability (some of them of living origin) but also complexes which, as for instance →chelation, comprise reversible reactions. In marine chemistry, theoretical knowledge is rapidly increasing to understand the equilibria of inorganic complexes of major and minor cations and anions in seawater.

compressibility of seawater reduction in volume due to pressure. Although the compressibility of fluids (including water) is low, it cannot be neglected at the high pressures that occur in the deep ocean.The compressibility is calculated from the →in situ temperature and salinity, and pressure, which again follows from the depth and the →density of the overlying water.

compressional wave see →P-wave.

concentration an amount of substance per unit of volume (or weight). Chemical concentrations were formerly often expressed as parts per thousand (ppt) (‰), parts per million (ppm) or parts per billion (ppb) being respectively grams, milligrams, or micrograms per 1000 g of seawater. When the amount of solute is expressed in grammoles per liter, one speaks of the molar or molal concentrations, dependent on whether the amount refers to a certain volume or weight. With marine organisms the concentration is usually denoted as the density of the species, which refers to the number of specimens per volume, per square meter (for →benthic organisms) or per square meter of the water column (for →pelagic species). The concentration is a very important quantity because it determines the probability that the solutes or the organisms meet each other. For chemical solutes this controls the reaction rates. With organisms the mutual interactions are, in a similar way, highly dependent on the probability that they meet each other.

condition factor, biological factor giving weight (W) relative to length (L) or diameter of an animal. In most animals body weight (W) shows a power relationship with linear body dimensions such as body length, height or diameter (L): $W = k\,L^b$. The power b usually has a value between 2.5 and 3.5 for different species, and can vary within one species for different life stages (larval, juvenile, adult, etc.), particularly when the body shape changes. When b is given a known fixed value for a species, the relative feeding or spawning condition of individuals can be described by a condition factor $K = W\,L^{-b}$, the factor being relatively high in wellfed or mature individuals and low in starved or postspawned individuals.

conductivity of seawater the electrical conductivity of seawater results from the ionized dissolved salts and is an accurate measure for the salinity. In fact, the routine determination of salinity by a →salinometer is done by measuring the conductivity.

confluence see →divergence.

connate water see →juvenile water.

conservation careful protection and planned management of a natural resource (ecosystem, fish stock, rare species) to prevent overexploitation, destruction or neglect.

consortium (microbiol.) an assemblage of different species of bacteria in which each organism benefits from the other.

constancy of seawater composition the relative abundances of major ions in seawater have a constant ratio to one another, hence to salinity (see →seawater). There are strong indications that the major ion composition of seawater has been constant over the past 400 million years.

consumers →heterotrophs all organisms that eat other organisms (either plants or animals). All animals, heterotrophic plants (such as →fungi), and bacteria are consumers. See also →producers.

contaminant any chemical substance, of natural origin (e.g., heavy metals, oil) occurring in augmented concentrations, or any chemical substance of anthropogenic origin present in the environment, but not necessarily related to noticeable biological effects.

continental crust see →crust.

continental displacement or continental drift, the movement of continents. In 1912 the German metereologist Alfred Wegener (1880–1930) promoted the theory that present-day continents were once joined in a supercontinent (→Pangea). See also →plate tectonics.

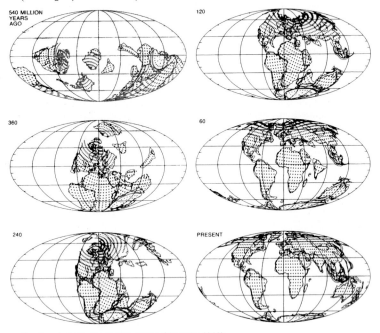

Continental displacement. (After Shepard 1963)

continental margin zone separating the emergent continents from the deep-sea floor. Where a continental margin results from a process of rifting and sea-floor spreading (see →plate tectonics) it is broad and gentle. Such margins are abandoned divergent →plate boundaries and are called passive; there is only limited tectonic activity. Examples of this type are the margins bordering the North Atlantic Ocean. An active continental margin contains a convergent plate boundary; it forms the leading edge of the overriding plate and parallels oceanic trenches. These narrow, steep-bounded margins are sites of intense earthquake activity with a volcanic arc on the continental side. The west coast of South America is a typical example.

continental shelf see →shelf sea.

continental slope the relative steep slope from the →offshore border of the continental shelf to the rise from the →abyssal plain. See also →deep-sea benthos.

continuity equation one of the basic equations in →hydrodynamics, formulating the condition of conservation of mass in a moving fluid.

continuous plankton recorder (CPR), or Hardy Plankton Recorder. A plankton sampler in 1925 invented by A.C. Hardy designed for towing behind commercial vessels, preserving the catch between two gauzes which are slowly wound together on a take-up spool enclosed in a formalin reservoir. The spool is driven by the outside water flow, turning a propellor at the back of the recorder. Operating routinely in the North Sea since 1932, and in the North Atlantic since 1948, the CPR catches form the best time series of plankton available. The Longhurst-Hardy Plankton Recorder is a

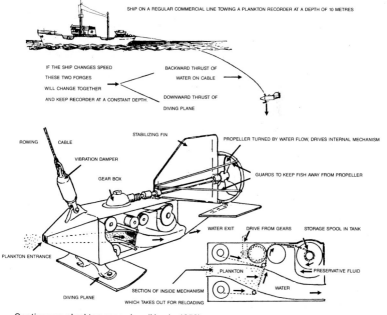

Continuous plankton recorder. (Hardy 1958)

version with electronically operated power pack and temperature-depth registration, to obtain discrete samples for tens of strata in one double oblique tow. See also →zooplankton samplers.

contour current bottom current along →bathymetric contours. The routes of flow of cold, dense water are mostly governed by the interaction of the →sea-floor topography and the forces produced by the rotation of the earth. This →Coriolis force pushes the currents towards the boundaries of the basins through which they flow, guiding currents along bathymetric contours. See also →contourite.

contourite deposited by →isobathic flows or →contour currents. Contourites consist of a sequence of thin beds with sharp contacts, are persistently laminated and well sorted and graded. They are usually found in sediments of the →continental slope.

convection vertical circulation in a gas or liquid under influence of instability.

convergence see →divergence.

convergent plate boundary see →plate boundary.

conversion factor figure used to convert an estimate expressed in a certain unit into another one, e.g., grams wet weight into grams carbon or calories into joules (see →caloric equivalents).

Cook, Captain James see →ocean exploration.

co-oscillating tide in a semi-enclosed sea area the tides are usually forced at the open side by the incoming oceanic tides. When the resonance period of the basin matches the tidal period, resonance may occur, giving very large amplitudes at the inner side of the area. One of the most famous resonance areas is the Bay of Fundy near Nova Scotia, with a tidal range of up to 17 m.

Copepoda (copepods) class of the subphylum →Crustacea (phylum →Arthropoda). The largest class of small crustaceans: over 7500 mainly marine species. Copepoda are small (only few species are larger than 1 mm) but extremely numerous and often dominate the →zooplankton. The body consists of six thoracic segments, of which one or two are fused with the head, forming the cephalothorax. The thoracic segments bear five pair of appendages (maxilliped and swimming limbs); the abdomen has five segments without limbs. The head has, stretched to the sides, two very large first antennae used for swimming, and one median naupliar eye. The feeding can be raptorial, or filtering of particles (mainly algae) from a water current effectuated by the anterior appendages. The eggs (in the order of 100 µm) are sometimes carried in egg sacs attached to the first abdominal segment, but generally released in the water, where they hatch in about a day. The larvae develop by successive molting through six naupliar (see →nauplius) and five →copepodite stages (I to V) to the sixth or adult stage. The generation time varies, depending on species, temperature and food supply,

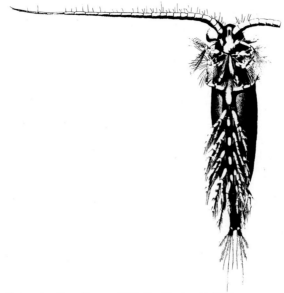

Copepoda. (From Barnes 1987 after Giesbrecht 1892)

from 2–3 weeks to 2–3 years. Sexes are separate; after fertilization the female stores the sperm and produces fertilized eggs over extended periods. For overwintering in temperate and artic regions, special resting stages (copepodite V) or winter eggs are produced in late summer. Copepods form a link in marine food chains between the →phytoplankton and →pelagic planktivore fish like herring.

copepodite larval stage of →Copepoda, with three pairs of functional swimming legs. There are five copepodite stages (CI-CV). See also →Copepoda.

coprophagy ingestion of feces. Especially the nonassimilated compounds in feces of →herbivores may have high nutritional value for →omnivores and →detritivores. See also →fecal pellets.

corals organisms belonging to the class →Anthozoa, phylum →Cnidaria. Although there are several orders within the Anthozoa, such as Antipatharia (black corals) and Gorgonacea (horny corals), which have skeletons, the term is in common use for the stony corals (→Madreporaria or Scleractinia). These animals possess a hard calcareous exoskeleton ($CaCO_3$). There are colonial and solitary forms. The bodies are anemone-like in shape and situated over the skeleton. The skeleton can be branched, massive, plate-like, etc. and is continually formed by the organism during its existence: coral growth. Of crucial importance in this calcification process are the Zooxan-

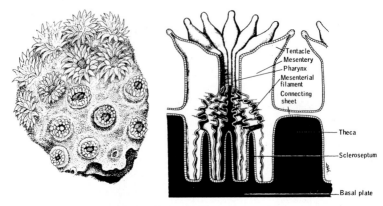

Corals. (From Barnes 1987 after Hyman 1940)

thellae, unicellular algae (e.g., *Symbiodinium microadriaticum*) living as symbionts (see →symbiosis) in the coral tissues and intimately involved in the coral metabolism. Corals are cosmopolitan but particularly abundant in tropical seas. They can build extensive structures: coral →reefs. On such reefs they function as the dominant organisms, in terms of shaping the habitat, in the complex coral reef communities. Large parts of reef systems are impoverished and destroyed by human activities including pollution.

co-range lines connect the points in a tidal chart where the tide has the same amplitude or tidal range (difference between high and low water level, or twice the amplitude).

core (1) see →earth's internal structure; (2) (oceanog.) vertical sample of the ocean bottom obtained with a →corer. Cores are especially useful in studying →stratification. A variety of oceanographic coring devices exist to extract cores from the sea bottom. See also →bottom sampling, corer, core drill.

core drill drilling device that cuts a →core from the drill hole. It is equipped with a hollow, cylindrical cutting tool and two nested tubes above it. The outer tube rotates with the drill and the inner tube receives and preserves a core from the material penetrated.

corer ocean-bottom sampler that is a hollow cylinder or box (see →box corer), lowered by cable and driven vertically in the sediment. After penetration, the corer contains a →core of the ocean bottom. Cylindrical corers vary in length of less than a meter to tens of meters, while the diameter varies from about 2.5 cm to more than 15 cm. **Gravity corer** penetrates the ocean floor solely by its own weight and velocity impact. It is

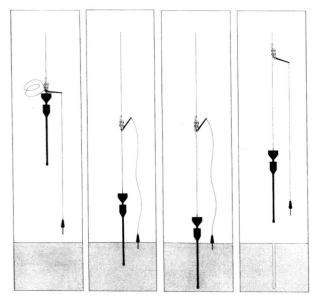

Piston corer

less efficient than a piston corer, which contains a piston inside the core tube. The piston remains on the seabed while the core tube falls past the piston. Hence, friction in taking the core is reduced. Giant piston corers may take cores of 20 m or more. The hydraulic piston corer is a type of piston corer which is lowered and retrieved by the drill string hanging from a drill ship. Thus, by repetitive operation in the same hole, cores are taken at about 4.5-m increments or more through unlithified intervals. The rapid punch of the cores is activated by the buildup of hydraulic pressures in special chambers. Sequences of 200 m or more can be cored in this manner.

Coriolis force fictitious force, described by the French mathematician G.G. Coriolis in 1835, that has to be introduced in the balance of forces when motions are considered with respect to a rotating coordinate system. The Coriolis force is introduced in the equations of motion if they refer to a coordinate system that is fixed on the rotating earth. As follows from the mathematical transformation from an absolute system to an earth-fixed system of coordinates, the Coriolis force has a horizontal component that is maximal at the poles and zero at the equator, and a vertical component that is maximal at the equator and zero at the poles. The latter can be neglected in →ocean dynamics, but in measuring the earth's gravity on a moving platform like a ship one has to apply a correction because of this effect

(Eötvös correction). The horizontal component depends on the sine of the geographic latitude and acts in the northern hemisphere to the right of a moving mass element and in the southern hemisphere to the left. In ocean dynamics the horizontal component of the Coriolis force is expressed as the product of the current velocity and the →Coriolis parameter.

Coriolis parameter the parameter giving the effect of the →Coriolis force on a current on the rotating earth is indicated with f and is a sine function of the geographic latitude. In small-scale dynamics this parameter is often assumed to be constant. This is called the f-plane approximation. The Coriolis parameter in that case is treated as if the earth's surface were a conical surface, tangential to the globe at the latitude studied. For larger-scale studies this approximation is often replaced by a parameter that changes linearly with latitude, with a proportionality factor β. This is called the beta-plane approximation.

correlation see →statistics and →stratigraphic correlation.

cosmogenic nuclides due to cosmic radiation on atmospheric gases and stratospheric dust, a number of natural radionuclides are formed, with various applications in ocean studies. Examples are (with their →half-lifes in parentheses) ^{7}Be (54 d), ^{10}Be (2.7 10^6 y), ^{14}C (5730 y), ^{26}Al (7 10^5 y) and ^{32}Si (170 y). Each of them has an application in marine research, depending on its half-life and geochemical behavior.

co-tidal lines connect the points in a tidal chart that have simultaneous high water. Where the tidal wave has the character of a progressive wave, the co-tidal lines give the crest of the wave at different time intervals. Where the tides have a standing wave character, the co-tidal lines are concentrated at the nodes. Where co-tidal lines for different instants converge, the nodes are concentrated in one point. These points are called amphidromic points. (Figure see p. 67)

Coulter counter trade name of electric type of particle counter. See also →particle counter.

counter current current running opposite to the main circulation, often compensating (part of) the water transport.

Cretaceous see →geological time scale.

Crinoidea sea lilies, class of the phylum →Echinodermata. They are the most primitive of the living echinoderms; the stalked Crinoidea were very common during the →Paleozoicum, now only found in deeper waters. Crinoidea have a five-sided symmetrical crown of arms bearing rows of pinnules, and therefore reminding of feathers. The sessile sea lilies are attached by a firm stalk to the substrate. Stalkless Crinoidea are free-moving feather-stars, occurring in shallow water. (Figure see p. 68)

critical depth depth at which different depth-dependent processes are in equilibrium. Specific in →photosynthesis: depth above which organic

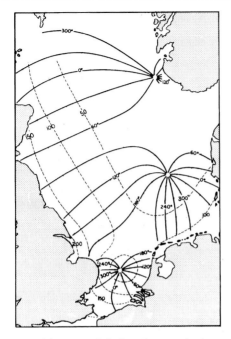

Co-tidal lines.
(After Proudman
and Doodson 1924)

material generated during photosynthesis exceeds the losses from all other
sources and phytoplankton population starts to grow.

cross-bedding a cross-bed is a single layer or sedimentation unit consisting of
internal laminae inclined to the principal surface of sedimentation. On the
basis of the character of the contact surface we distinguish planar cross-
bedding, with more or less tabular to wedge-shaped units, and trough cross-
bedding, in which bounding surfaces are curved. Small-scale cross-bedding
(maximum thickness of individual units not exceeding 4 cm) is mostly the
result of the migration of small current and wave →ripples. Large-scale

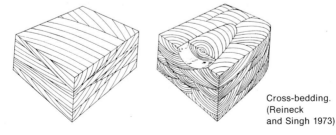

Cross-bedding.
(Reineck
and Singh 1973)

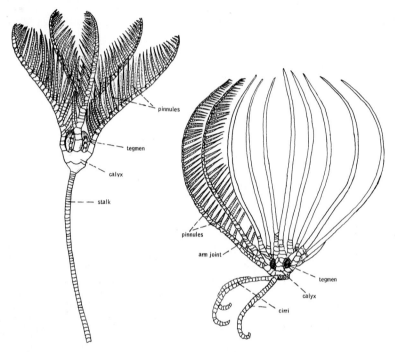

Crinoidea. (From Barnes 1987 after Hyman 1955)

cross-bedding is caused by the migration of →megaripples, longshore bars, and others; units may reach thicknesses of 1 to 2 m. See also →sedimentary structures.

crust outermost shell of the earth, overlying the mantle above the →Mohorovičič discontinuity. The definition includes the overlying sediments. The crust can be divided into the part constituting the continents (continental crust and the part forming the floor of the ocean's basins (oceanic crust). Continental crust is composed of granitic rocks and denser (probably gabbroic) rocks at depth. It is generally 30 km thick, but may be over 60 km thick under mountain regions. Oceanic crust is only 5–10 km thick and is composed of a layer of →basaltic rocks (underlain by more peridotitic rocks at depth). See also →earth's internal structure.

Crustacea subphylum of the phylum →Arthropoda. Crabs, hermit crabs, shrimps, and lobsters together with smaller and less familiar arthropods form some 42 000 species of crustaceans. Although the species differ considerably in size and shape, they are all variations on a rather strict

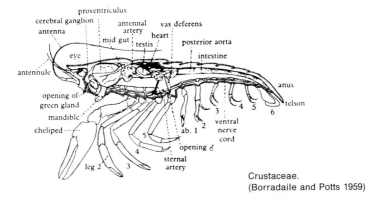

Crustaceae.
(Borradaile and Potts 1959)

design of the body, which consists of somites (segments), that can bear one pair of appendages each. In all crustaceans the head is formed by the first six somites, of which the first has no appendages (or limbs). The second somite bears the antennules, the third the larger antennae. The three other somites of the head come after the mouth, and bear the mandibles, maxillules and maxillae, respectively; all small limbs that fit together round the mouth and move the food into it. The trunk is less uniform than the head, with, depending on the species, varying numbers of somites and appendages that are modified for different functions. The first part of the trunk is the thorax, built of 6–11 somites, with appendages that are the actual limbs. In →copepods and →amphipods, the first thoracic limbs are accessory maxillae, hence called maxillipeds; in →decapods there are three pairs of maxillipeds. The thoracic limbs can be very large with claws (chelipeds) or modified for swimming, digging, crawling, carrying eggs, etc. The number of remaining (abdominal) somites is variable.

cryophile see →psychrophile.

Cryptomonadida see →Cryptophyceae.

Cryptophyceae class of the division →Cryptophyta (see →algae) containing the nanoplanktonic flagellate genus *Cryptomonas* and closely related organisms (also called Cryptomonads). They contain →chlorophyll-c, phycocyanin, phycoerythrin, alloxanthin, α-→carotene instead of β-carotene. The carbohydrate storage product, starch, is found outside the plastid. The plastid color ranges from blue to red, but is mostly chestnut. Conspicuous ejectosomes produce kinked strands on discharge. The Cryptophyceae are protists (see →Protozoa), which can be classified as an order of the →Mastigophora. Examples: *Cryptomonas*, a species usually abundant in estuarine environments, but often overlooked because discharge of the ejectosomes easily destroys the cells; *Chilomonas*, a common colorless genus in polluted water). Chryptophyceae are protists (see →Protozoa), which can also be listed as the order Chryptomonadida of the phylum →Mastigophora.

CTD recorder stands for Conductivity-Temperature-Depth recorder. The instrument measures the temperature, electrical conductivity, and underwater pressure electrically, and from the latter two the salinity and depth are computed. The advantage of this instrument over the older techniques with →Nansen bottles and →reversing thermometers is that it gives the data directly with a very high vertical resolution, although →calibration with the older instruments stays necessary.

ctenidium (pl. ctenidia) ciliated organ of →Bivalvia, consisting of many filaments, and suspended in the mantle cavity. The ctenidia (also called gills, because of their resemblance to the gills of fish) are mainly filters for sieving food particles from the water, and are to a lesser extent involved in the oxygenation of blood in the gill veins. The ctenidia effectuate by ciliary movement (see →cilium) a water current from the inhalant opening or siphon, through the ctenidia, towards the exhalant opening or siphon. The suspended particles retained by the ciliated filament are transported over the ctenidium surface towards the mouth lobes (or →labial palps), which pass the selected fraction to the mouth for ingestion. The remaining, not ingested material is removed as pseudofeces.

Ctenophora (comb-jellies) subphylum of the phylum →Cnidaria. A small group of marine species, in principle with the same body structure as →medusae with →ectoderm, →endoderm, →mesogloea, and gastro-vascular cavity. The body is generally spherical, with eight rows of combs with →cilia, by the movements of which the comb-jellies swim. Two branched, and adhesive tentacles, which are used to collect small planktonic prey animals like →copepods, can be retracted into tentacular sheaths in the sides. Tentacles in a 1-cm-wide comb-jelly can be 50 cm long. In some species the tentacles are reduced and the prey is taken directly from the water with the mouth lobes. Ctenophores are divided into the classes Tentaculata (e.g., *Pleurobrachia*, a cosmopolitan species especially abundant in coastal waters, and the open-sea species *Mnemiopsis* and *Bolinopsis*) and Nuda (no tentacles, e.g., *Beroe*, a comb-jelly feeding exclusively on *Pleurobrachia*).

Ctenostomata order of the class Gymnolaemata (phylum →Bryozoa). The body walls are not calcified. Example: *Alcyonidium*.

culture techniques the complex of technical measures for the cultivation and propagation of living organisms or tissues. This comprises optimalization of physical and chemical requirements such as tank size, temperature, light, substrate, food supply, and prevention of diseases and contamination.

Cumacea (cumaceans) order of class →Malacostraca, subphylum →Crustacea (phylum →Arthropoda). Small crustaceans (up to a few cm) that live

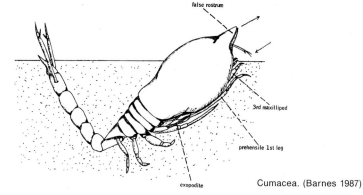

Cumacea. (Barnes 1987)

buried in sand and mud bottoms, with the head sticking out of the sediment.
→Thorax and head are greatly enlarged, which is typical of the cumaceans.
They feed by filtering particles from the water, or directly from the bottom.
Cumaceans are periodically →pelagic; swarming is related with reproduc-
tion and →molting.

curie unit of radioactivity $(1 \, Ci = 3.7 \, 10^{10}$ desintegrations per second)
equivalent to the radiation by 1 g of radium, the approximate quantity of
radium isolated by Madame M. Curie (1867–1934) in her professional
lifetime. The curie has been replaced by the →becquerel.

current meter instrument to measure speed and direction of water currents.
Different types of current meters have been used in oceanography. Before the
introduction of automatic self-recording instruments, measurements were
mainly made from ships, with often interference of ship's movements. Modern
instruments are anchored on a line with subsurface floats, which makes the
possibility of surface movements interfering with the measurements much
smaller. They record the current data on tape, and can remain on station for
many months. Recovering of deeply moored instruments is often by
→acoustic releases. The current sensor is usually mechanical (impellor or
rotor) and the number of revolutions per unit time is proportional to the
current speed. The direction is measured by recording the position of a current vane with respect to a compass. As the first measurement gives a time
average and the other a momentaneous value, caution is required in interpreting the results. A further problem may arise from →fouling when instruments
are deployed too long. Other sensors, such as the →Doppler current meter,
are acoustic or electromagnetic and have a high time resolution.

current ripple see →ripple.

cuticula (cuticle) superficial, noncellular layer covering certain organism,
secreted by the hypodermis. See also →exoskeleton.

Cyanobacteria (syn. Cyanophyceae or blue-green algae) →prokaryotes like →bacteria. Their cell nucleus has no envelope, in contrast to the →eukaryotic true algae, and they lack plastid membranes and mitochondria. →Chlorophyll-a, β-→carotene and phycobilin are their pigments for →photosynthesis. Many species owe their characteristic blue-green color to phycocyanin. Some cyanobacteria, however, are more yellowish green and some are even red due to phycoerythrin. There are four major subgroups of cyanobacteria. First the chroococcacean cyanobacteria, which are unicellular rods or cocci. They reproduce either by binary fission or by budding. Second, the pleurocapsalean, which are single cells enclosed in a fibrous layer. The third group contains the oscillatory cyanobacteria, which form filamentous structures, exclusively composed of vegetative cells, known as trichomes. The fourth group exists of the heterocystous cyanobacteria, which form differentiated cells known as heterocysts, when growing in the absence of fixed forms of nitrogen. In →picoplankton the chroococcacean *Synechococcus* plays a significant role in the →primary production of the oceans.

Cyanophyceae see →Cyanobacteria.

cycle (1) a time interval (day, year, tide, etc.) during which a sequence of recurrent (successions of) events is completed; (2) a course of a series of events that recur regularly and often lead to the starting point: →carbon cycle, nutrient cycle, →life cycle, →biogeochemical cycle.

cyclonic flow under the influence of the →Coriolis force on earth, water flowing around a low pressure area according to the →geostrophic balance (so with the omission of other, usually smaller scale factors) rotates in a cyclonic sense. Because of the differences in the direction of the Coriolis force in both hemispheres, this cyclonic flow is curved to the left in the northern hemisphere and to the right in the south. The opposite of cyclonic flow is anticyclonic flow, around an area of high pressure.

Cyclostomata (1) eel-shaped cartilageous fish (lampreys), see →Pisces; (2) order of the class Stenolaemata (phylum →Bryozoa). The →lophophore is circular without an →epistome. There are neither →avicularia nor →vibracula. They have various types of →trochosphere larva. Many recent and fossil species, e.g., *Crisia*.

cypris larva see →Cirripedia.

cyst a →resting stage formed by some bacteria, algae and protozoa in which the whole cell is surrounded by a protective layer. N.B. cyst should not be confused with →spore, which is formed inside a cell.

D

Darwin, Charles Robert see →evolution.

data assimilation see →models.

dating see →absolute age.

datum level see →biohorizon.

DDT →polychlor pesticide, first synthesized in 1874 and its →insecticidal
properties discovered in 1939. Has saved hundreds of thousands of human
lives, e.g., by reducing populations of insects carrying diseases such as
malaria. Persistence of DDT and its adverse effects on wildlife led DDT to
be restricted severely as from the start of 1970. The technical mixture
dichlorodiphenyltrichloroethane contains three isomers, of which p,p'-DDT
is the principal component; the others are the o,p'-DDT and p,p'-DDE, the
minor constituent to the technical mixture. In the environment DDT is
rapidly converted to p,p'-DDE (in biota) and to p,p'-DDD in the abiotic
compartments (sediment and water).

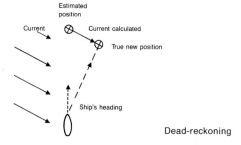

p,p'-DDT   p,p'-DDE   DDT

dead-reckoning by regularly comparing the actual position of a ship with the
position inferred from its course and its speed in the water since the last
position, the mean current experienced during this time interval is

Dead-reckoning

estimated. This information is still the basis for many ocean current charts. This method of current measuring is typically a →Lagrangian representation.

debris see →detritus.

debris flow see →sediment-gravity flow.

Decapoda (decapods) order of the class →Malacostraca, subphylum →Crustacea (phylum →Arthropoda), containing all the familiar crabs, shrimps, crayfish, and lobsters, together some 10 000 species (for general description of body see →Crustacea). The first three pairs of the eight pairs of thoracic appendages changed to maxillipeds, leaving in total ten appendages to function as legs, hence "decapoda" (ten legs). The decapods used to be divided into the suborders Natantia (→pelagic: shrimps) and Reptantia (→benthic: crabs, lobsters, etc). At present they are divided into two suborders: Dendrobranchiata (shrimps that do not carry the eggs on pleopods, example: *Penaeus*) and Pleocyemata. The Pleocyemata are divided into a number of infraorders, containing the different types of crustaceans that carry the eggs on pleopods till they hatch. These are the Stenopodidea and Caridea (shrimps like *Crangon*), the Astacidea (crayfish and lobsters), the Palinura (spiny lobsters) and the Brachyura (crabs).

Decapoda. (Barnes 1987)

decomposition the return of constituents of organic matter to ecological cycles by the activities of living organisms (decomposers). See also →carbon cycle.

deep scattering layer (DSL) see →scattering layer.

deep-sea benthos →benthos living in and on the bottom of the sea deeper than about 200 m, i.e., deeper than the →continental shelf. For its food it is largely dependent on organic matter from the →euphotic zone. Therefore it is not surprising that →abundance and biomass of the benthos decline with increasing water depth. However, diversity (species richness) is mostly relatively high in the deep sea. In general, the deep-sea benthos is divided into three vertical zoogeographical zones: bathyal (200–2000 m, →continental slope), abyssal (ca. 2000–6000 m, continental rise and deep-sea plains and hadal (6000 m and more, trenches and troughs).

Deep-Sea Drilling Project (DSDP) long-range project for drilling in deep water (down to 6000 m) for scientific purposes. The technique for drilling in deep waters was not feasible until the 1960s. The main problem was to maintain the same position above the bore hole. In shallow water this is accomplished by anchoring or raising a platform on legs. In deep water these solutions are impractical and a dynamic positioning system was developed. This system uses acoustic beacons dropped on the sea-floor near the bore hole. The signals are picked up by hydrophones in the ship's hull and computers determine and control the direction and amount of movement necessary to keep the drilling vessel within certain limits of the desired position. Under the auspices of →JOIDES the project started in 1968 with R.V. *Glomar Challenger* and since then over 800 holes have been drilled, some of them up to the lower portion of the basaltic layer. See →layer 1, layer 2, etc.

deep-sea fan terrigenous, core- or fan-shaped deposit located off the shelf in front of large rivers and →submarine canyons, of which it often forms an extension. The sediment is supplied by rivers or by longshore transport over the shelf, usually over the inner shelf or near-shore. The sediment is funneled to the fan by way of a canyon; the dominant transport mode is a →turbidity current. Sediment is deposited where the current reaches the lower gradients of the ocean basin. Deep-sea fans receive sediment from specific hinterland areas as do alluvial fans on land. The volume of many fans, in relation to sediment supply, indicates that they are millions of years old. A striking example is the Ganges cone (2500 km long, terminating at a depth of ca. 5000 m). Smaller fans occur along most continental slopes. Most, if not all, fans are marked by one or more deep-sea channels, which are usually the distal extension of a →submarine canyon. At the Ganges and Indus cones the fan reaches almost to the shelf break but the canyons are short, whereas the Zaire (Congo) canyon begins in the river estuary, crosses the shelf, and extends to the base of the continental slope where the Zaire deep sea begins at ca. 3500 m depth.

deep-sea sediments sediments found on the deep-sea floor. Except for some →turbidites, these sediments are often →pelagic in origin. Pelagic sediments are those derived from the precipitation of dissolved or suspended matter in the open ocean together with the accumulation of the remains of →planktonic organisms. These sediments are generally found at depths greater than 1000 m and comprise →oozes and →red clays. Pelagic sediments may be classified as eupelagic or hemipelagic, according to the abundance of terrigenous or volcanic material. Eupelagic sediments are deposits in which less than 25 % of the fraction coarser than fine silt is of terrigenous or volcanic origin. Such sediments usually form far from the continents. Hemipelagic sediments contain more than 25 % terrigenous or coarse volcanic-derived material. These deep-sea sediments usually accumulate near the continental margin, so that continentally derived sediment is more abundant than in eupelagic sediments.

deep-water species animals which live in and are confined to the deep sea. Their mostly →stenothermal character, i.e., the absence of an ability to deal with fluctuating temperatures, prevents their existence in the warmer upper water layers, but in polar regions some can be found at shallow depths.

deep water wave see →short wave.

deformation radius a length scale, also called Rossby radius of deformation, at which the effects of the rotation of the earth cannot be neglected, compared with the gravitational restoring forces. It equals the (long-)wave velocity divided by the →Coriolis parameter. It is of the order of 100 to 1000 km for surface waves (depending on the sea depth), but it can become smaller (order 10 km) for baroclinic →internal waves, where the internal phase speed is the pertinent velocity scale.

delta accumulation of river-supplied sediment deposited at the coast where a stream decelerates by entering a large body of water. The stream may be of any size, and the outflow may be into an ocean gulf, a lagoon, an estuary or a lake. The term "delta" orginated with Herodotus (5th century B.C.), who described the triangular plain of river alluvium at the Nile river mouth, broadly resembling the Greek capital letter delta (Δ). Other types of deltas that have no triangular form are also called deltas, the best-known being the Mississippi river delta, which has a "bird-foot" configuration with extended lobes and digitate tributaries. Deltas show a large variety of morphological and depositional features, reflecting variations in river discharge regime, coastal (marine) energy conditions, and geological structure: deltas are complexes of river mouths, distributary channels, flats, bays, swamps, beaches, beach ridges, and dunes. The delta plain is the subaereal part of the delta. Usually a lower delta plain can be distinguished, extending inland to the limit of tidal influence; further landward is the upper delta plain. The subaqueous part can be divided into a basal prodelta, consisting of more or less horizontal layers of primarily fine silts and clays deposited from suspension, and a delta front of coarser sediment including river mouth bar sands, tidal ridges, and beach deposits.

demersal organisms bottom-dwelling aquatic organisms, e.g., demersal fish, in contrast to →pelagic organisms, which live in the water column.

Demospongiae class of the phylum Porifera or sponges, with a skeleton of siliceous spicules which are never six-rayed, or spicules composed of silica and spongin (horny fibers) or both. Flagellated chambers are small and rounded and have small →choanocytes. Sponges possess jelly. Most sponges belong to this dominant and widely distributed class; they occur both in freshwater and in the sea.

denitrification the conversion of nitrate by bacteria into nitrogen gas, resulting in the loss of nitrogen from ecosystems. This process occurs mainly in →anaerobic systems but is also possible in →aerobic systems.

density (1) (phys.) the mass per unit of volume (symbol ϱ). In older literature this physical property is also called specific weight. The density of seawater has been determined empirically as a function of temperature, salinity and pressure, the →equation of state. This empirical relation is given as a polynomial expression. Since the density of seawater is only slightly larger than 1000 kg m^{-3}, one often uses the so-called density excess, $\sigma = \varrho - 1000 \text{ kg m}^{-3}$. In older literature the density anomaly σ was used, being $\sigma = (\varrho - 1) 1000$, whereby ϱ was given in g cm^3 (cgs units), using the now obsolete equation of state of seawater according to Knudsen; (2) (biol.) see →abundance and →concentration.

deoxyribonucleic acid (DNA) a polymer of nucleotides connected via a phosphatede-oxyribose sugar backbone; it is the genetic material of the living cell.

DNA

deposit-feeding feeding on bottom material with the food value being determined by its organic content. Particles may be swallowed indiscriminately or sorted for size prior to ingestion.

desalination the process of removing salts from seawater or sea sediment. It is accomplished by (vacuum) distillation or by high pressure, reverse osmosis for large-scale production of potable water in oceanic regions with insufficient freshwater resources.

desorption the reverse process of →sorption, which itself can be divided into adsorption and absorption (see →sorption). The speed of desorption is generally different from that of adsorption (rapid) and absorption (slow). This is because the adsorption layer will quickly establish a new equilibrium between the sorbed material on the solid and that in the solution. Consequently, the release of the absorbed material will determine the desorption speed. For practical use the term half-time of desorption (time of 50 % loss) gives a qualitative indication of the strength of the sorption.

detergents group of alkali-organic compounds with surface-tension reductive properties, applied for cleaning solid surfaces and cleansing oil-contaminated waters. Detergents with, if possible, nontoxic properties are applied to solubilize oil slicks and thus increase the bacterial decomposition of the pollutants. Various compounds are poisonous in concentrations of over 0.1 mg dm^{-3}. This is mainly because the change in the surface tension of water influences the transfer of substances in and out of organisms.

detrital sediments (geol.) sediments consisting of fragments of pre-existing rocks, produced by weathering and erosion, and in general transported from another location. See also →sediment classification.

detritivores organisms feeding on →detritus.

detritus (1) (biol.) remains (debris) of dead organisms, more or less decomposed so that the origin cannot easily be traced. It is often mixed with fine inorganic material. If it originates from plants, it is called phytodetritus. Animals, microbial organisms, or fungi may consume detritus and are then called detritivores; (2) (geol.) a collective term for loose rock and mineral material (debris) produced by the desintegration and weathering of rocks and removed from its site of origin; esp. fragmental material, such as sand, silt, and clay.

Devonian see →geological time scale.

diagenesis post-depositional changes of sediments as a result of interacting physical, chemical and biological processes.

diapir a flow structure resulting from the upward intrusion of a less dense rock mass through overlying denser rock. Salt and shale are the most common rocks involved in diapirs. Intrusive rocks can also form diapir-like features but the term is usually restricted to plastic flow.

diarrhetic shellfish poisoning (DSP), a transitory gastro-intestinal disorder following consumption of →shellfish contaminated by lipid-soluble toxins. The toxins are concentrated mostly in the hepatopancreas of mollusks and their origin has been traced to planktonic →dinoflagellates ingested by the filter-feeding shellfish prior to the time of harvest. Members of the genus *Dinophysis* have been identified as producers of these toxins. This kind of DSP has been reported in several coastal areas worldwide. See also →paralytic shellfish poisoning.

diatoms see →Bacillariophyceae.

diel see →diurnal.

diffluence see →divergence.

diffusion the transport across a unit surface can be quantified by putting it proportional to the concentration gradient. The proportionality factor is called the diffusion coefficient. For molecular transport processes the diffusion coefficient can be taken independently of the scale length of the transport, for turbulent transport the coefficient of eddy diffusion depends on the scale length. This is the normal condition in ocean waters. In sediments where the interstitial diffusion at the other hand is molecular, it should be corrected for the hindrance by sediment particles (→tortuosity). For substances sorbed to sediment particles, the apparent diffusion coefficient can be calculated by dividing the molecular diffusion coefficient (of the order of $5 \ 10^{-6} \ cm^2 \ s^{-1}$) by the →distribution coefficient K_d, supposing that a kind of equilibrium may occur between substances dissolved in pore water and substances sorbed to sediment particles.

digestion the enzymatic breakdown of food substances to molecules which can be absorbed and used in cellular processes of living organisms. The degradation of large ingested molecules takes place by the cleavage of chemical bonds, almost without exception by →hydrolysis, the addition of a water molecule across the bonds.

digestive enzymes organic catalytic substances functioning in the breakdown of food substances by cleaving molecular bonds in the substrate. Digestive enzymes are very specific, acting only on certain substrates. So, there are many carbohydrases (carbohydrate-digesting enzymes), proteases (proteolytic enzymes) and lipases (fat-degrading enzymes). They work best outside the cell, for their optimum pHs lie either on the acidic or basic side, while the cell interior constantly requires an almost neutral pH (about 7.4).

dimethyl sulfide (DMS), the most abundant volatile sulfur compound in seawater. In most cases DMS is of biogenic origin with as precursor β-dimethyl sulfoniopoprionate (DMSP): $(CH_3)_2S + CH_2COO^-$. In the air DMS undergoes photochemical oxidation to sulfate and thus contributes to the low pH of rain. On a yearly basis the global marine sulfur flux is in the range of 30 to $170 \ 10^{12} \ g \ S \ y^{-1}$, which is in the same order of magnitude as anthropogenic fluxes from fossil fuel burning.

Dinoflagellata (dinoflagellates), see →Dinophyceae.

Dinophyceae (syn. Peridinia, dinoflagellates) class of unicellular algae containing the Dinoflagellata and some related groups. Their size is ca.

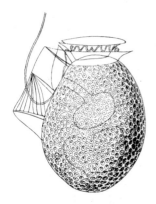

Dinophyceae. (Schütt 1895)

5 μm to 2 mm. Typical are the two corresponding grooves. The Dynophyceae have a relatively large nucleus (dinokaryon), with clearly recognizable chromosomes also during the interphase. The dinoflagellates are protista (see →Protozoa) that from a zoological point of view are not algae, but →Protozoa and an order (Dinoflagellidae or Pyrrophyta) of the subphylum →Mastigophora (phylum Sarcomastigophora). →Photosynthetic forms contain chlorophyll-c and peridinin. They store starch outside the plastids. Cell walls, if present, consist of cellulose lined by sporopollenin. The walls of →resting spores are wholly made up of sporopollenin. →Phagotrophy, ecto-, and endo-→parasitism, →bioluminescence, and toxic →blooms are found in this class. Corals have dinoflagellate endosymbionts. The colorless *Protoperidinium* and *Noctiluca*, and the pigmented *Gonyaulax* and *Ceratium* are important in marine phytoplankton. Fossil resting spores are found in sediments from the Silurian onwards.

discoasters star- or rosette-shaped calcareous →nanofossils of still unknown origin probably related to →coccoliths. Widely used for zonation of paleogene sediments and for the determination of the →Pliocene-Pleistocene boundary, because of their extinction at the end of the Pliocene.

dispersion (1) (phys.) term used in oceanography in two different ways: in relation to the propagation of waves and in relation to the spreading of a property in different directions. In the first case it is applied to waves of different frequency starting at the same location, or to the spreading of some local disturbance that can be considered to be composed of such waves. After some time the waves with the lower frequency are getting ahead because of their higher →group velocity. The relation between frequency and group velocity, or, more properly, between frequency and →wave number is therefore called the dispersion relation. In the second case dispersion means the spreading of some dissolved or suspended substance because of a combination of →shear and turbulent →diffusion at right angles; (2) (biol.) the distribution pattern in a plant or animal population, see →distribution.

displacement experiments experiments in which mobile →benthic animals are displaced from their original habitat to another, in order to study differences in various aspects of the behavior and →bioenergetics of these animals in a new versus a well-known surrounding; also used with birds in studying migration mechanisms.

displacement volume rough measure of weight, used in plankton field work. The plankton catch, or a size fraction of it obtained by sieving over a certain →mesh size, is dried on a filter by suction, and put in a known volume of seawater. The volume increase gives the displacement volume, and, assuming a specific weight of roughly 1, an indication of the fresh weight.

dissolved organic matter →see organic carbon.

distal situated away from; especially from place of attachment; opposite of →proximal.

distribution (1) the spatial →dispersion or spatial pattern of a population. Three classes of patterns are often distinguished: regular (there is a fixed distance between the individuals of a population, which might be a consequence of competition), random (each individual may occur anywhere in an area with equal probability; so it is not affected by the occurrence of other individuals) and contagious (individuals occur in clusters or patches; within clusters their distribution may be regular or random). Most populations in the marine environment show a contagious distribution. (2) When a large number of measurements or counts of some variable is collected, the observations can be arranged in numerical order and grouped into classes. The set of frequencies of each class show the observed distribution along the axis of measurement. Such a distribution can be summarized by a "measure of central tendency" like the →mean, →median and →mode, and by a "measure of dispersion", like the →standard deviation and the →variance. The way in which values of a variable are distributed can be theoretically described by a probability distribution. Commonly applied distributions are the binomial distribution, Poisson distribution, and normal distribution. The bell-shaped normal distribution is the most widely used distribution in →statistics and is completely determined, if its parameters, the mean, and the variance are known.

distribution coefficient the empirical factor for describing the partition or sorption-desorption equilibria of dissolved and particulate substances; symbol K_d: is defined as the ratio of the concentration in particulate or solid matter, for instance expressed in $\mu g\ g^{-1}$, and the concentration in solution (also in the same units). This K_d is applicable for transport-, budget- and hydrodynamic modeling, as well as for calculating apparent →diffusion coefficients of transport in bottom sediments. Sometimes K_ds are expressed in the dimension ml g^{-1}, which differs a factor equal to the solid matter density from the dimensionless K_d factor.

diurnal (syn. diel) processes or rhythms related to the diurnal rotation of the earth, more in particular to the natural light and dark periods, either physically, like the diel temperature cycle with heating and cooling of the surface layer at low latitudes, or physiologically, like →photosynthesis in algae in the light, or behaviorally, like nocturnal feeding in animals. The latter is often coupled with diurnal →vertical migration, the ascent of zooplankton and nekton to surface water at dusk, and descent at dawn. This is clearest in echosound recordings which occur at →deep scattering layers (DSL's) at depths of several hundred meters by day, rising to and diffusing in the upper 100 m at night. See also →diurnal tide.

diurnal inequality of tides difference between successive tidal ranges in a semi-diurnal tide. See also →harmonic analysis of tides.

Diurnal inequality of tides

diurnal tide tide with high water and low water only once in the course of (approximately) one day, as for instance in southeast Asian waters. See also →harmonic analysis.

divergence/convergence (phys.) the expansion/contraction of a transported property like mass or heat. In the →continuity equation the divergence of mass is zero. Often divergence is related to a source and convergence to a sink. In a two-dimensional plane like the sea surface, the current may show a divergent or convergent pattern, but as this brings about vertical motion, and thus no three-dimensional divergence or convergence, the terms diffluence or confluence should be preferred.

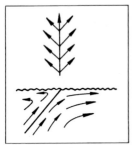

Divergence. (After Scharnow 1978)

divergent plate boundary see →plate boundary.

divers (biol.) animals that catch their food by diving, e.g., whales, some seabirds.

diversity term used for the degree to which the total number N of individual organisms in a given ecosystem, area, →community or →trophic level is divided evenly over different species. Diversity is minimal when all organisms are of the same species, as, e.g., in a monoculture or (almost) in intensive agriculture. Diversity is maximal in stable natural environments with a maximal variation in substrate and life conditions (or, where the number of →niches is maximal), as, e.g., on coral reefs in the tropical oceans. Diversity can be expressed quantatively by various diversity indices. The simplest is: diversity $d = S/\log^2 N$, where S is the total number of species. The Margalef index $d = (S-1)/\ln N$ is a modification, with minimum value $d = 0$ for a monoculture. More complicated indices of diversity are used when it is also taken into account that there are different →distributions possible of the total number of individuals N over the number of species S, or in other words, when the concepts evenness or dominance are introduced (an even distribution means maximal evenness, while the opposite is maximal dominance of the single species making the whole system). When, moreover, the concept richness (the number of species among a fixed number of individuals or per unit surface) is incorporated, the best way to express diversity is by the Shannon and Weaver index: $H' = -\Sigma p_i \log^2 p_i$, where $p_i = n_i/N$ and n_i the number species i.

diversity index see →diversity, biological.

diverticulum blind-ending tubular or sac-like protuberance from a cavity.

DMS see →dimethyl sulfide.

DNA see →deoxyribonucleic acid.

DOC dissolved organic carbon, see →carbon, organic.

doldrums climatic zone in the tropics. See →climate.

dolomite see →carbonate, mineral.

dolphins see →marine mammals.

DOM dissolved organic matter, see →carbon, organic.

domestic waste see →waste.

dominance see →diversity.

Doppler effect the apparent change in frequency of a wave signal from a source that is moving relative to the detector, theoretically described in 1842 by C. Doppler (1803–1853). In measuring instruments a moving source is not directly used, but rather the backscattered signals from a source that is mounted near the detector. In oceanographic instrumentation this effect is used with acoustic signals, and with electromagnetic signals (light and radar). Acoustic applications are the Doppler Sonar to determine the speed of a ship with respect to the bottom, and the acoustic Doppler current meter. This latter has the advantage that the flow is not affected by the instrument, as is the case with propellor-type current meters. In the laser Doppler turbulence meter, laser light is used to measure velocity fluctuations in a volume of less than a cubic millimeter. Radar signals, backscattered from surface waves, are used to measure surface currents by this principle.

dorsal situated at, or relatively nearer to, the back, i.e., the side of a bilateral animal which is normally turned away from the ground or substratum; opposite of →ventral.

doubling time the time needed for a population to double in size.

downwelling the downward movement of surface water under influence of the wind, because of →Ekman transport; opposite of →upwelling.

drag coefficient a nondimensional factor in the relation between the square of the flow velocity at a given distance from a boundary and the surface stress. It depends on the turbulent character of the →boundary layer and is, under certain conditions, considered constant.

dredge sampling gear for collecting materials from the seabed as →benthic flora and fauna, stones or minerals etc. The two types most commonly used are the pipe dredge and the frame dredge. The pipe dredge is a metal cylinder of about 2 m with a diameter of about 50 cm. The sharpened front edge eventually cuts the rock and the rear has a grating with enough space to allow fine sediment to pass through. The frame dredge has a bag attached to a metal frame. The frame dredge covers more ground than the pipe dredge, but is far more likely to get caught under ledges that cannot be fractured. See also →bottom sampling. (Figure see p. 85)

drift the transport of organisms or objects by water currents. When for organisms this transport is independent of the (swimming) activity, as in the case of drift of eggs and larvae, the term passive drift is used. See also →sediment drift.

drift current current that is directly driven by wind stress. The classical theoretical solution for the structure of a drift current is given by the →Ekman spiral solution. Empirical relations between wind and current that have been established for different regions are often strongly influenced by local conditions and by the quality of the current and wind data used. They usually result in a →wind factor that gives a (mostly linear) relation between wind and drift current.

drift net see →nets.

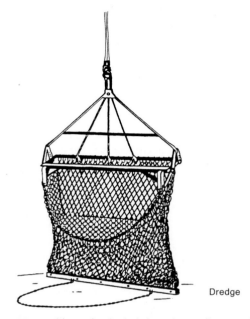

Dredge

drilling making a circular hole in rock or sediments with a drill or other cutting tool. In the ordinary →rotary drilling method the material from the bore hole is crushed, and some characteristics of the undisturbed rock section penetrated can be obtained afterwards by →well logging. If one wants to preserve a core from the bore hole, a →core drill is used. See also →bottom sampling.

DSDP see →Deep-Sea Drilling Project.

DSP see →diarrhetic shellfish poisoning.

dumping see →ocean dumping.

Dunbar line dividing line in the North Atlantic Ocean between the cool-temperate and subarctic zones; a zoogeographical term often used in taxonomic literature.

dune (1) (subaerial) low mound, ridge, bank or hill of loose windblown granular material (sand, volcanic ash, often mixed with shell fragments), either bare or covered with vegetation; retains a characteristic shape even when moving over considerable distances. See →coast; (2) (subaquatic) large sand ripple, see →megaripple.

dust (geol.) dry, solid matter consisting of clay- and silt-size particles, so finely divided that they can be easily picked up and carried over long distances in suspension in the atmosphere. Sources of marine dust include, e.g.: deserts, volcanic eruptions, salt spray from the seas, and extraterrestrial meteoric dust. Eolian dust often also comprises sand-size particles. It is activated by whirlwinds originating from strong convection above the intensely heated surface of deserts. See also →eolian sediment.

dynamic topography chart of vertical variations of the sea surface, computed with respect to a certain reference level, and based upon observations of the depth distribution of temperature and salinity. From these observations the specific volume of the water can be calculated, and this again gives the height of the sea level above a certain reference level. If this reference level were horizontal, the real topography would be obtained, but in practice at best only an approximation is obtained.

dynamics see →ocean dynamics.

E

earthquake (seismic event) sudden motion or trembling in the earth caused by
volcanic activity or by an abrupt movement of rock masses along fault
planes that results from instantaneous release of slowly accumulated elastic
strain. Earthquakes often lie along the interface of →plates, where one
plate is overriding another. See also →plate tectonics.

earth's internal structure the →crust is the outermost shell of the earth,
consisting of relatively light materials. The →Mohorovičić discontinuity
marks the boundary between crust and the denser underlying mantle. The
mantle extends to a depth of roughly 2900 km and is thought to consist of
ferromagnesian silicate minerals such as olivine and pyroxene. At a depth of
ca. 700 km a transition occurs to a denser phase, dividing the mantle into
upper and lower mantle. The mantle essentially behaves as a solid, although
the asthenosphere marks a weak approximately 200-km-thick zone where
it may be partially molten. The asthenosphere develops at 50 to 250 km
beneath the earth's surface and defines the boundary of the rigid outer
→lithosphere which includes the →crust and part of the upper mantle.
Presumably, →plates of the lithosphere, move over the asthenosphere, (see
→plate tectonics). The mantle below the asthenosphere is called mesosphere. A central core of high-density matter is distinguished from the less
dense peripheral mantle, the surface separating them being known as the

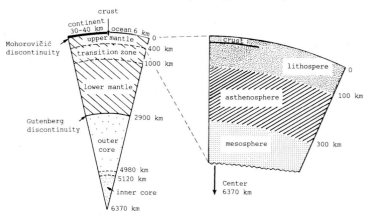

Earth's internal structure

Gutenberg discontinuity. The core probably consists of an iron-nickel alloy and is divisible into a liquid outer core of about 2200 km in radius and a solid inner core about 1300 km in radius.

easterlies climatic zone with predominant easterly winds. See →climate.

ebb tide period in the tidal cycle between high water and the next low water. The ebb current is the current during the falling of the sea level. In the case of a →progressive tidal →wave, the reversals of the current take place just between the times of high and low water, and the ebb current cannot well be defined.

ecdysis see →molting.

Echinodermata (echinoderms) phylum containing starfish, brittle-stars, sea urchins, sea lilies and sea cucumbers. The echinoderms ("spiny skins") are exclusively marine, most of them →benthic, and have the following characteristics: the body has a radial symmetry (in principle five sided); they have an internal skeleton formed by calcareous plates (ossicles) in the skin, which may articulate with each other (like in starfish and brittle-stars) or which are sutured together into a stiff skeleton (sea urchins). Usually the skeletal plates bear spines giving the echinoderms the typical spiny appearance. An important feature is the branched system of coelomic canals running through the body known as the water-vascular system. This system of canals is filled with watery fluids; water can enter the system through a sieve-like plate on the surface of the animal, called the madreporite. The system has surface appendages, the often suckered tube feet, moved by hydraulic pressure. There are no →nephridia or other special excretory organs; the nervous system is simple and attached to the epidermis. The →coelom is spacious, the digestive tract well developed. Most echinoderms live on or in the seabed, where they are →deposit feeders (sea urchins like *Echinocardium*), →carnivores (like the starfish *Asterias*) or collecting

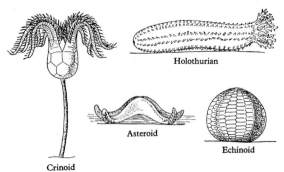

Holothurian

Asteroid

Echinoid

Crinoid

Echinodermata. (Borradaile and Potts 1959)

particles from the water (sea lilies). Eggs and sperm are released in the water, and from the fertilized eggs hatch the typical echinoderm larvae, which develop into plankton (see →meroplankton). The phylum is divided into the class Stelleroidea (with subclasses →Asteroidea and →Ophiuroidea), the class →Echinoidea, the class →Holothuroidea, and the class →Crinoidea.

Echinoidea class of the phylum →Echinodermata. Free-moving echinoderms, sea urchins, heart urchins, and sand dollars. Spherical or sometimes flattened body without arms. The ossicles (or calcareous plates in the skin) are flat and sutured completely into the typical echinoid skeletal case or test; other ossicles developed into movable spines that cover the body. Besides such spines, the echinoids also have podia or tube feet (see →Echinodermata), and use both for locomotion on hard substrate like rocks, to crawl on seaweeds, or to bury and move through sandy or muddy bottoms. The regular sea urchins have a typical structure with dents round the mouth for scraping food from the substrate, called the lantern of Aristotle. They feed on plant material and →detritus. The irregular heart urchins, which live in soft bottoms, eat more or less unsorted sediment containing algae, detritus, bacteria, etc. Echinoids are found in all seas and oceans, and are also an important group of fossils.

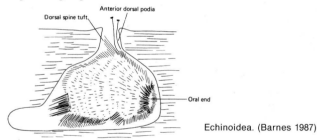

Echinoidea. (Barnes 1987)

Echiura small phylum of marine worms with a cylindrical trunck and large →proboscis. They live in burrows in sandy and muddy sediments, mainly in shallow water. (Figure see p. 90)

echogram print-out of a device that emits sound pulses and scans a distant body through recording sound echoes. Echography is used in medical and marine science. See also →echosounding.

echolocation (syn. sound scanning) locates obstacles by emitting sound and receiving and interpreting the echoes. Several species of bats and toothed whales (including dolphins) have developed this system to a high degree of perfection, while some other mammals, such as shrews, rats, seals, and penguins seem to use echoes to a lesser extent. The physical properties of water allow sound to be propagated much more effectively than in air.

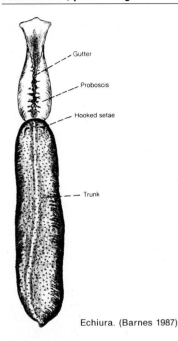

Echiura. (Barnes 1987)

Echolocation has been found in more than ten species of toothed whales. The sounds used are in the ultrasonic range (from 0.25 to 220 kHz). There is growing evidence that the skulls of the marine echolocating species contain several properties for emitting the sound in a narrow forward beam. Pulse frequency increases when the predator closes in on its prey, both in whales and bats.

echosounder, penetrating see →seismic instruments.

echosounding originally, a method of measuring the water depth. The depth is measured by the travel time of acoustic signals, consisting of brief pulses of sound between 20 and 200 kHz to the sea floor and back. Because of differences in temperature and salinity the acoustic velocity differs regionally. →Scattering layers consisting of organisms may sometimes give echoes that can be taken for bottom echoes (phantom bottom). In marine geology, lower frequencies (<7 kHz) with higher energetic content are applied. Then, not only the sea bottom reflects the sound pulses, but also layers of different density deeper in the sediment. The →echogram then gives information on the sediment structure. See also →seismic instruments.

ecological efficiency the fraction of the energy consumed at a given →trophic level that is transferred to the next higher trophic level.

ecological models simplified representations of the interactions of organisms or assemblies of organisms with the environment. Ecological models of complete natural systems, or →ecosystems, are called ecosystem models. In models of an aquatic ecosystem, the system is broken down into a series of components, such as →phytoplankton, zooplankton, detritus, nutrients, etc.; and numbers (system variables) are used to describe the state of each component (say, the biomass) at a given time. In ecosystem models, as in other models, their use as a tool or as a means for prediction depends on properties like accuracy, realism, and general manageability, that can be mutually conflicting.

ecological monitoring see →monitoring.

ecology study of the interactions between organisms and their →environment. Autecology concerns a single species organized in one or more populations, while synecology applies to studies of populations of more than one species organized in a community.

ecosystem strictly speaking any part of the →biosphere can be considered an ecosystem, comprising living and nonliving constituents with a complicated network of mutual interactions, and as a whole having inputs from and outputs to the surrounding biosphere. Those parts of the biosphere, whose internal interactions are much stronger than the interactions with the outer world, might be considered ecosystems, like ponds, inland seas, or sea areas (that are isolated in terms of, e.g., temperature, salinity). In more complicated cases, like, e.g., estuaries, a coastal ecosystem, or the →benthic and →pelagic ecosystem, the specific character is due to the specific physical nature of the area, while interactions with the outside biosphere may dominate internal dynamics. The clearest example of the latter may be an ecosystem that is a section of a river or a sea area with strong currents.

ecotoxicology the science dealing with the identification and quantification of the biogeochemical processes and →toxic effects of chemical substances occurring in enhanced concentrations and physical agents introduced into the environment by human activities. Effects on living organisms are studied, especially on populations and communities within defined ecosystems; studies include the transfer pathways of those agents and their interactions with the environment.

ecotypes morphological and/or physiological variants of a species that occur and survive as a distinct group in a specific kind of habitat. These variants may be genetic races or merely physiological acclimation.

ectoderm superficial germ layer of an animal embryo, developing mainly into →epidermis, nervous tissue and →nephridia (when present). The term is applied to the germ layer while it is still a demarcated region of the embryo, after gastrulation but before differentiation into the derived tissues. All the tissues at later stages derived from the embryonic layer are ectodermal.

Ectoprocta see →Bryozoa.

eddy a more or less circular movement of water elements. Eddies are an important mechanism in nonadvective transport, and, depending on their relative size, are part of the mixing process (see →turbulence). The scale of eddies varies from diffusion scales (cm or less) to hundreds of kilometers. The transport of dissolved substances or heat by turbulent processes is called eddy diffusion (see also →diffusion) and transport of momentum by turbulent processes is called eddy viscosity.

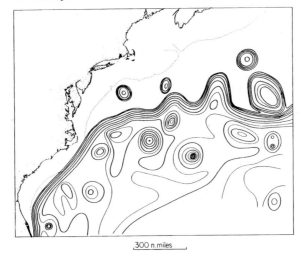

Eddies. (After Gill 1982)

300 n.miles

edge effects disturbances of homogeneous environmental conditions, caused by the vicinity of natural or artificial obstacles: boulders at the seabed, the glass walls of an aquarium, the hull of a ship. Edge effects make experiments with →pelagic organisms very difficult.

edge wave a wave that is trapped near a coast with a sloping beach. The rays of this wave are gradually refracted towards the coast. For a given coastal angle of incidence there will be a line at some distance along the coast where the reflected rays run parallel to the coastline. Inshore of this line there is a pattern of standing (trapped) waves and offshore of it the amplitude will decay exponentially. See also →trapped waves.

eel grass see →sea grasses.

EEZ see →Exclusive Economic Zone.

Ekman layer the surface layer over which in the case of the →Ekman spiral, the current becomes reduced to $1/e$ (where e is the ground number of the natural logarithm, 2.71828...) of the surface value. The most important part of the →drift current takes place in this layer. An Ekman layer is also present as a →boundary layer that forms the transition between a →geostrophic current and the near-bottom layer where the bottom friction is the main factor.

Ekman number a nondimensional number, E, that gives the ratio between the →frictional force per unit mass to the →Coriolis force. It is given by: $E = K/fL^2$, with K the coefficient of →eddy viscosity, f the →Coriolis parameter and L a typical length scale. When the Ekman number is small, frictional forces can be neglected.

Ekman spiral a solution of the equations of motion (see →ocean dynamics) for the current driven by a homogeneous →wind stress was in 1905 given by the Swedish oceanographer V.W. Ekman (1874–1954), in order to explain the angle between wind and ice drift as observed by Nansen on his drift through the North Polar Sea with the *Fram*. A condition is that there are no horizontal pressure gradients (sea surface slopes) and in the most elementary case it is further assumed that the sea is very deep (no influence of bottom friction) and thus that the wind stress and the →Coriolis force are the only external forces. In large, deep oceans these conditions are approximated. The result is a current that makes at the surface an angle with the wind direction; to the right in the northern hemisphere, to the left in the southern, and that with increasing depth decreases in strength and further

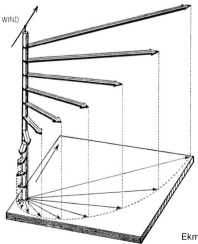

Ekman spiral. (After McLellan 1965)

turns away from the wind direction. The point of the current vector thus describes a spatial spiral, the Ekman spiral. For other cases (shallow seas) comparable solutions can be obtained. The theory of Ekman is important in understanding processes like →Ekman transport and →upwelling.

Ekman transport the total wind-driven transport of water in the Ekman solution of the equations of motion is found to be at right angles to the direction of the wind. This result can be found by vertical integration of the depth-dependent currents as described by the Ekman spiral. In the case of large-scale cyclonic wind field (see →cyclonic flow) it follows that there is →divergence of water at the surface, which has to be replaced by upwelling water (Ekman suction). Similarly anticyclonic wind fields give downwelling of surface waters (Ekman pumping).

Elasmobranchii see →Pisces.

electrodes conducting surface used for the determination of solute concentrations. The functioning of reversible electrodes is based on Nernst's Law (electrical potential proportional to the logarithm of the solute concentration). Electrical potentials are measured relative to a reference electrode. Reversible electrodes exist for the determination of →pH, →oxygen and alkaline-, and alkaline-earth metals. At polarographic electrode surfaces the solute is removed by electrochemical reaction. This results in an electrical current. The maximum conversion rate is limited by the rate at which the solute is transported to the electrode. The current output is linearly proportional to the solute concentration. Polarographic electrodes exist for the measurement of concentrations of oxygen, nitrous oxide, trace metals and electrochemically reactive organic substances.

electron-transport-phosphorylation the synthesis of →ATP involving a membrane-associated electron transport chain and the creation of a proton-motive force.

elements our natural environment is built up of 90 different chemical elements, numbers 1 through 92, omitting numbers 43 (Tc or Technetium) and 61 (Pm or Promethium), of which only man-made, relatively short lived radio isotopes now occur in the environment. The transuranic elements 93 through 103, all in the Actinide series, also are man-made only. All 90 natural elements exist in seawater in concentrations ranging (when excluding the H and O in the H_2O water molecule) from about 0.47 moles per liter for Na (sodium) down to the femtomolar range (10^{-15} moles per liter) for trace elements like Eu (Europium), Pd (Palladium), or Bi (Bismuth).

El Niño the name given by the inhabitants of the west coast of S. America to the phenomenon that around Christmas time (hence the name "the child") warm equatorial water moves southward and replaces the colder waters that are present there for the rest of the year. It appears that at 3- to 7-year intervals this phenomenon is much more pronounced and gives rise to exceptional climatic conditions. The atmosphere-ocean interaction that is held to be the reason for this interannual oscillation is part of the so-called →Southern Oscillation. Strong El Niño years upset the ecological conditions along large parts of the South American coast, with effects on the fisheries.

emigration process of migration of individuals of a population out of an area. As opposed to →immigration, migration towards, into an area.

endobenthos see →endofauna and →benthos.

endoderm germ layer of animal embryo, composed of cells which have moved from the surface of the embryo into its interior during gastrulation; developing into the greater part of the gut with its associated glands. The term is applied to the germ layer while it is still a demarcated region of the embryo, after gastrulation but before differentiation into derived tissues. All tissues at later stages derived from the embryonic layer are called endodermal tissues.

endofauna (syn. infauna) organisms that live buried in the sediment for the greater part of their lives, such as most worms and bivalves. Sampling methods (see →bottom samplers) for endofauna include cores and grabs, of which the contents are washed over a sieve in order to remove the sediment; opposite of →epifauna.

endospore bacterial spore formed within the cell; it is extremely resistant to heat as well as to other harmful agents.

endostyle organ for ciliary feeding in, e.g., →Urochordata. It consists of a ciliated and glandular groove or pocket in the ventral wall of the →pharynx. It produces threads of →mucus to which food particles adhere that are passed backwards by ciliary action.

energy cascade in general the motions in the sea cover a wide range of frequencies, from gradually varying currents to small-scale rapidly changing turbulent →eddies. The distribution of kinetic energy over this frequency range can be presented as an energy spectrum of ocean motions. In this spectrum, energy is exchanged between different frequency bands. Most energy is fed in the large-scale motions by the →air-sea exchange and the →tides, and the energy is dissipated in the small-scale motions where it is irreversibly converted into heat (see Kolmogoroff →inertial subrange). The flow of energy from large to small scales is indicated by the term "energy cascade".

energy from the sea the sea takes part in the energy cycle of the earth by accumulating and transporting energy received from the sun, the atmosphere, and the tides, and dissipating it or returning it to the atmosphere or to outer space. Various methods have been proposed to extract a small fraction of this energy for human use. The main possible sources are thermal, wave, and tidal energy. In the first case, the temperature difference of the water at different depths is used to drive a thermal engine (Ocean Thermal Energy Conversion or OTEC). In the second case, the kinetic energy of the waves is used and the tidal energy is extracted by using the potential energy of the water in a basin filled at high tide. Under present conditions energy from the sea is only economically feasible under special circumstances. Tidal energy is used in places with a large tidal range (e.g., in the Rance estuary of Brittany in France).

enrichment ratio is in most cases equal to the →distribution coefficient or concentration factor; the first term is usually applied for sorption to sedimentary particles, the second used in biological accumulation.

Enteropneusta see →Hemichordata.

enthalpy see →thermodynamics of seawater.

entrainment (1) (geol.) the process in which substances are incorporated in solid matter during its formation; (2) (phys.) one-directional movement of water through a boundary surface between two bodies of water with different properties such as a →pycnocline. The entrained water mixes with the water across the boundary, which thus changes its properties, while the water on the other side of the boundary keeps its original properties. Entrainment therefore differs from mixing, where the two water masses exchange properties. In case of entrainment there should be either advection or a gradual shift of the place of the boundary surface.

entropy thermodynamical quantity: the energy (in joules) absorbed by a reversible process divided by the absolute temperature (in Kelvin).

environment the sum of conditions and objects (living and nonliving) that act upon organisms or groups of organisms and determine their form and/or functioning.

environmental monitoring see →monitoring.

environmental stress see →stress.

Eocene see →geological time scale.

eolian sediment (or aeolian sediment) after Aeolus, the Greek God of the winds. Wind-blown sediment of volcanic (volcanic ash) or nonvolcanic (sand, desert dust) origin that often is transported over large distances and eventually settles on land, e.g., dunes or in the ocean. Well known is the reddish-brown quartz originating from weathering of continental rocks. Other terrigenous minerals, such as →calcite and biogenic components, notably →phytoliths, freshwater diatoms and even fungus spores also occur in atmospheric dust. Fine particles are carried as "aerosols" over the ocean; upon deposition and vertical settling through the oceanic water column the sediment accumulates on the sea floor at rates in the order of about one millimeter per thousand years. Such red clays, undiluted by marine biogenic (calcite, →opal) or →turbidite inputs only occur in restricted areas of the sea floor, remote from the continents. Common components of eolian sediments may form the clay minerals →chlorite, →illite, →kaolinite and montmorillonite (→smectite). See also →dust.

Eötvös correction see →Coriolis force.

ephyra pre-stage of →medusa.

epibenthos benthic organisms living on the surface of the benthos. See also →benthos.

epicenter the location on the earth's surface below which an →earthquake originates. Compare →hypocenter.

epidermis outermost tissue layer of an animal, in invertebrates only one cell-layer thick.

epifauna bottom animals that live on the surface of the sea floor, such as most crabs, shrimps and starfish; opposite of →endofauna.

epipelagic community →community of species and organisms living in the epipelagic zone (0–200 m, see →zonation).

epiphytes organisms living attached to plants, e.g., epiphytic algae on sea grasses, Bryozoa, Foraminifera.

episodic of or pertaining to, or having the nature of, an episode; incidental, occasional.

epistome small, movable, lip-like lobe overhanging the mouth (in some →Bryozoa).

epizoa animals living attached to other organisms, without being parasitic.

equation of state equation giving the relation between temperature, salinity and pressure of seawater. As the required accuracy has increased since the first determination of this equation for seawater by M. Knudsen in 1901, it has recently been redefined. See →thermodynamics of seawater.

equatorial currents the equatorial zone is dynamically one of the most important parts of the ocean. The equatorial currents flow in a westerly direction in a zone of about 1000 to 1500 km width, as parts of the subtropical →gyres, under the influence of the →trade winds. In the zone of low winds between the trade wind belts there is an eastward countercurrent, which for most of the year is positioned north of the equator. In the Indian Ocean there is a seasonal reversal in this system north of the equator, because of the reversing →monsoon winds. In the equatorial zone also eastward subsurface countercurrents have been observed, which are sometimes indicated as the Cromwell Current (Pacific) and the Lomonosov Current (Atlantic).

equatorial wave at the equator the →Coriolis force changes sign and this makes it possible for westward propagating planetary waves and zonally traveling long waves to be trapped in a zone on both sides of the equator. Equatorial planetary waves are thought to play a role in the dynamics of the equatorial countercurrent and of the →El Niño system.

erosion loosening, dissolution, or wearing away of the Earth's crust and transportation by natural agencies. The term is sometimes restricted by excluding transportation or weathering.

erratic (geol.) transported rock fragment, deposited at some distance from its place of origin and generally resting on bedrock or sediment of different lithology. The term is generally applied to fragments transported by glaciers or floating ice (see →glaciomarine sediments and →ice rafting). Erratics range in size from pebbles to house-size blocks and occur either free or as part of a sediment.

erythrocruorin high molecular weight extracellular →hemoglobin molecules, occurring, for instance, in →Annelida.

esophagus (oesophagus) part of the alimentary canal between →pharynx and stomach, passing the food along by peristaltic contraction.

estuarine zone the estuarine part of a sea area. See →estuary.

estuary a semi-enclosed coastal body of water that has a free connection with the open sea and within which seawater is mixed with freshwater from land. The important physical processes are the effects of vertical →stratification because of the salinity differences and the longitudinal and vertical mixing by →turbulence. This turbulence may be caused by the tides, but also by current →shear. The landward limit has been defined at 0.01 ‰ chlorinity. Estuaries are an unstable environment because the mixing of the so-called endmembers fresh and salt water is variable. Estuaries are classified by their origin (coastal plain estuaries, lagoonal estuaries, fjords, faultblock estuaries), by the type of mixing (saltwedge, partially mixed, and well-mixed estuaries), and by the tidal range (microtidal: tidal range 0–2 m; mesotidal: tidal range 2–4 m, macrotidal: tidal range more than 4 m). Mixing is largely dominated by the interaction of river outflow and the tides, although wind forcing may temporarily dominate the estuarine circulation. Saltwedge estuaries have a surface layer of low salinity water flowing out over, and being well separated from a bottom layer of seawater. Vertical mixing is slow, the tidal range is usually small. Well-mixed estuaries have a strong vertical mixing, the tides dominate, and the morphology of the estuary allows the tidal wave to enter far into the river mouth. Partially mixed estuaries are intermediate, with some salinity stratification but no sharp boundary between low salinity surface water and high salinity bottom water. Estuarine sediments may show a large variation of sands, silts, muds, and shell or other carbonate deposits. The sediment can be supplied by rivers as bottom load or suspended load, or from coastal seas by inflow along the bottom. Less important sources are shore erosion, and organisms.

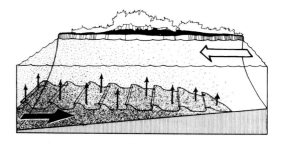

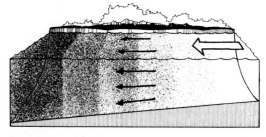

Estuary. *top*: stratified estuary; *below*: mixed estuary (Pritchard 1955)

Waste discharges from population centers and industries may be locally or regionally important. Sedimentation is often highly irregular: during exceptional floods or storms more sediment may be deposited in a few days than during many years of normal conditions. In tidal estuaries usually a turbidity maximum develops, with high suspended matter concentrations, at the point where the inward flow along the bottom meets the outward river flow, and low flow velocities occur near to the bottom. The location of this point varies with spring and neap tides and with river discharge, so that a zone of muddy sediments and high suspended matter concentrations develops at the head of the estuary. Many estuaries act as sediment traps but the delicate balance between sediment supply, deposition, and outflow into the coastal sea, in combination with the large variability, makes it difficult to estimate whether there is export or import of sediment in an estuary. Some estuaries, like the Thames estuary, have existed for many thousands of years at approximately the same location. In such estuaries the sediment deposition has compensated for the rise in sea level that occurred during the →Holocene.

eu- prefix meaning well, good.

Eucarida superorder of the class →Malacostraca (subphylum →Crustaceae, phylum →Arthropoda), containing the larger Malacostraca with stalked eyes and highly developed →carapace fused with all thoracic segments, e.g., →Euphausiacea (krill) and →Decapoda (shrimps, lobsters and, crabs).

Euglenophyceae see →Euglenida.

Euglenida (syn. Euglenophyceae) order of the subphylum →Mastigophora (phylum Sarcomastigophora, see →Protozoa), containing elongated flagellates with numerous green →chromatophores, or colorless. They have a gullet with a →flagellum and a contractile vacuole with a reservoir opening into it. The cells can contain paramylum and sometimes oil as reserves. There is no transversal groove. They have a very elastic outer covering and specialized protoplasmic contractile fibers which permit contraction and elongation of the body in a characteristic squirming called euglenoid movement. There are many freshwater forms and some marine. Some colorless species devour prey larger than themselves. Other colorless species take only dissolved substances or are endoparasites in copepod eggs. →Photosynthetic euglenoid flagellates are found in estuarine environments and on coastal mudflats.

eukaryotes all organisms, except bacteria and blue-green algae (see →prokaryotes) are eukaryotes. They have a distinct nuclear membrane surrounding the nucleus with the chromosomes.

Eulamellibranchiata types of →Bivalvia (or Lamellibranchia) in which the →ctenidia are greatly enlarged and used for filtering food from the water. The junctions between the gill filaments are permanent and vascular. The foot is well developed, the →byssal glands are small or absent. They have one or two adductor muscles. Examples: *Ostrea, Cerastoderma.*

Eulerian representation representation of the motion of a fluid by noting the flow at different points in the current field. Current meters at fixed positions give an Eulerian representation. See →Euler-Lagrange transformation.

Euler-Lagrange transformation the velocity distribution in a fluid can be represented in two different ways: by following the fluid elements and noting their velocity and acceleration at any time and place (→Lagrangian representation), and by giving the flow in different points of the current field and its development in time as local accelerations (→Eulerian representation). In principle, the transformation from one data set to the other is only exact if the current were known at any time and location. The problem of the Euler-Lagrange transformation is posed by the practical situation that this condition is never completely realized. These two representations are named after the respectively German and French mathematicians L. Euler (1707–1783) and L. Lagrange (1736–1813), who contributed to the hydrodynamic theory.

euneuston see →neuston.

eupelagic sediments see →deep-sea sediments.

Euphausiacea (euphausiids) order of the class →Malacostraca, subphylum →Crustacea (phylum →Arthropoda). Shrimp-like →pelagic crustaceans. They occur in all the oceans in almost all regions. They are small (up to a few cm) and differ from the →Decapoda in, e.g., the thoracic appendages being all similar (no →maxillipeds). They are →suspension feeders, filtering water through long fringes on the appendages, and eat algae but also zooplankton. They can occur in enormous swarms and provide the major part of the food of baleen whales and some fish. The most highly differentiated genus is *Euphausia* (krill) with 30 species.

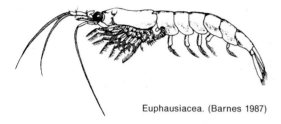

Euphausiacea. (Barnes 1987)

euphotic zone depth to which 1 % of surface illumination penetrates. See also →light.

eury- prefix meaning wide.

euryhaline able to tolerate a wide range of salinities and thus a wide variation in osmotic pressure of the environment; opposite of →stenohaline.

euryoecious refers to species which a wide tolerance to fluctuations in environmental factors, e.g., →eurythermal, →euryhaline, etc; opposite of →stenoecious.

euryphagous thriving on all kinds of food material, including carrion; opposite of →stenophagous.

eurythermal or eurythermous, able to tolerate wide variations in ambient temperatures; opposite of →stenothermal.

eustasy the worldwide sea-level regime and its fluctuations. The origin of these fluctuations is commonly attributed to absolute changes in the volume of seawater, e.g., by fluctuations in the amount of continental ice or by the thermal expansion. Eustatic movements may also be caused by tectonic movements of sea floors or land masses. See also →sea level and →isostasy.

eutrophic regions highly productive areas due to a large natural or man-induced nutrient load (→upwelling areas, →estuaries) but with a relatively low species →diversity. Eutrophic is usually used rather loosely and not well defined with regard to nutrients and productivity level. The same holds for oligotrophic (low productivity and nutrient content) and mesotrophic (in between eutrophic and oligotrophic).

eutrophication enrichment of natural waters with inorganic nutrients (ammonia, nitrate, phosphate) by which phytoplankton growth is stimulated. Eutrophication leads to increased biomass, →decomposition, in the worst cases resulting in oxygen depletion and →mass mortality. An important effect of eutrophication is therefore, next to increased productivity in the area, an increased risk of a complete system deterioration. In most cases the enhanched possibilities for phytoplankton growth lead to shifts in the species composition of the algae, which becomes dominated by types that can absorb the supplied nutrients most quickly. Phosphate and nitrogen euthrophication leads to a shift in the system from diatoms (which need also silicate) to flagellates. Food chains change accordingly, and the eutrophicated area becomes unsuitable for many species that lived there before.

evaporation transition of matter from the liquid into gaseous phase. Dissolved substances remain in higher concentration in the solution. In seawater evaporation increases the salinity. Evaporation from the sea surface is promoted by a low humidity of the air and by strong winds. The energy required for evaporation (→latent heat) is extracted from the water that cools down. High evaporation is found in the subtropics, where in addition precipitation is low. Here the highest salinities of the surface water are found.

evaporite sedimentary rock composed of minerals that have been precipitated from solutions concentrated by the evaporation of the solvent. Examples include: gypsum, anhydrite, rock salt and various nitrates and borates. Evaporites usually form under arid or semi-arid conditions on coastal plains just above normal high-tide level or in semi-isolated marine basins with replenishment of seawater over a restrictive entrance. Evaporite deposits can reach considerable thicknesses, with minerals layered in a sequence depending upon their relative solubilities.

evenness see →diversity.

evolution the development of living organisms from the most primitive →prokaryotes to vertebrates and man. The fossil record of marine and terrestrial sediments strongly suggests an oceanic genesis and that the invasion of the land took place only a short (geological) time ago. Representatives of most phyla can be traced in marine sediments while on land and in terrestrial sediments only a limited number of phyla are represented. About 3.5 billion year ago →photoautotrophic bacteria appeared. After the emergence of →Cyanobacteria, equipped with a →photosynthetic apparatus capable of decomposing water and simultaneously releasing oxygen, the ocean and the atmosphere changed from an anoxic state to the oxic state. This permitted aerobic metabolism to evolve, leading ultimately (about 1.5 billion years ago) to the emergence of

→eukaryotes. The proliferation of eukaryotic species was associated with further changes in the composition of the atmosphere and the hydrosphere. The concept of evolution was combined with the idea of natural selection by C.R. Darwin (1808–1882) and A.R. Wallace (1823–1913). Especially the work of the former is famous. His voyage as a naturalist on HMS *Beagle* (1831–1836) laid the foundation for his theory of the origin of species by means of natural selection published in abridged form in 1859 and further elaborated on in books on variation of animals and plants under domestication (1868) and the descent of man (1870). He was an all-round naturalist who published, among other things, on subjects such as coral reefs (1842) and the geology of volcanic islands (1844). See also →biostratigraphy.

Exclusive Economic Zone (EEZ) legal concept introduced by →UNCLOS III. It can now be said to constitute a rule of international custom. The EEZ is an area of the sea beyond and adjacent to the →territorial sea. Its breadth is not to extend beyond 200 →nautical miles from the →baselines, from which the breadth of the territorial sea is measured. The coastal state enjoys certain sovereign rights over the EEZ, namely: rights for the purpose of exploration, exploitation, conservation, and management of all natural resources of the seabed, its subsoil and of the overlying waters. The coastal state also enjoys jurisdiction with regard to the establishment and use of artificial islands, marine scientific research, and the protection of the marine environment.

Exclusive Fisheries Zone (EFZ) zone, recognized by international customary law, in which states may claim exclusive fishing rights up to 200 miles (including the →territorial sea). Most states have entered into agreements permitting them to fish (at a certain charge) in the EFZ of the other contracting state. The EFZ is generally taken as part of the →Exclusive Economic Zone (EEZ).

excretion see →excretory organs, →exudation of organic matter and →animal excretion.

excretory organs organs that are specialized in the removal of metabolic waste products. The kidney, as it occurs in higher animals, is absent in invertebrate species. Here, many types of duct-like structures occur with a similar function (e.g., nephroducts in →Annelida, antennal glands in →Crustacea), which have in common: (a) a large surface area; (b) a well-developed blood supply; and (c) an excretory pore which ends either in the digestive tract or on the outside of the body wall.

exoskeleton the distinguishing feature of the →Arthropoda, where the →cuticle consists of chitinous material, sometimes containing calcium carbonate and calcium phosphate (→Crustacea), which is excreted by the hypodermis, the skin layer directly under the epidermis.

expeditions in oceanology, research carried out at open sea using ocean-going research vessels. A great number of regional and worldwide national and multinational expeditions have increased the knowledge of the world oceans considerably. At present such large-scale expeditions are also prepared through the Scientific Committee of Ocean Research (SCOR), of the ICSU (International Council of Scientific Unions) and the International Oceanographic Committee (IOC) located at Unesco. The objectives are (through multi-vessel expeditions) to study predetermined subjects in predetermined ocean regions.

exploration of the ocean see →ocean exploration.

exposure monitoring see →monitoring.

external digestion digestion outside the organism. Microorganisms often produce extracellular enzymes and some animals bring digestive enzymes outside themselves on their prey. As a result macromolecules are split to smaller subunits, which can be easily taken up by the organism.

extinction (1) (biol.) the disappearance of certain animal or plant species in the course of →evolution. Such extinctions have certainly something to do with changing environmental conditions during geological time. Extinctions in the past were not randomly distributed but largely confined to short periods. On land the extinction of the large reptiles at the end of the Cretaceous is the best known. See also →mass mortality, →Iridium-rich level, →microfossils, and →biohorizon; (2) (phys.) generally the decrease of energy by dissipation in a beam of radiating energy: light, sound, or mechanical energy (waves). Scattering of energy from the beam is not considered extinction, but it also contributes to the energy decrease in the direct beam. Extinction of light is also called absorption (see →optics in the sea).

extracellular metabolites of algae see →exudation of organic matter.

extraction (trace metals) the transfer of substances into another liquid medium. The technique of extraction can be applied to concentrate small amounts of trace elements from seawater. Problems which can occur in the determination of trace elements in seawater with atomic absorption spectrophotometry are very low trace metal concentrations (nM) and high salt matrix. The method is based on the →complexation of trace elements with complex formers viz. ammonium pyrrolydine dithiocarbonate (APDC) and diethylammonium diethyldithiocarbonate (DDDC). These complexes are extracted into an organic solvent, viz. freon or methyl-isobutylketone (MIBK). After extraction, the complexes are destroyed with nitric acid and the metals back extracted into double quartz-distilled water. The advantages of the method are: (a) the salt matrix is removed, and (b) the elements are concentrated by a factor of 50 or more.

exudation of organic matter (sometimes called excretion) the escape of
soluble organic compounds from all kinds of water plants (macrophytes,
phytoplankton) during →photosynthesis. Normally less than 5 % of the C
fixed during photosynthesis is exudated, but higher figures have been
reported at high light intensities and at the end of phytoplankton blooms.
Some are probably artifacts. This fraction does not form an important
contribution to the →DOC pool in sea, because the compounds concern
small amounts of →amino acids and larger amounts of short-chain acids
(e.g., glycolic acid), glycerol, carbohydrates and polysaccharides which are
rapidly assimilated by bacteria. Exudation may be passive as well as active.
Active exudation includes extracellular metabolites of algae-like metal-
→chelating compounds in metal-polluted waters and antibacterial compounds.

F

facies the appearance and characteristics of a rock unit, usually reflecting the environmental conditions of its origin without regard to age. e.g.: A sedimentary facies is the sum total of the lithological (lithofacies) and biological (biofacies) characteristics of a sedimentary deposit. Moreover, a sedimentary facies analysis reveals a picture of the ancient sedimentary environment, including the nature of the materials (→detrital, ionic), the paleogeographic position of the area of deposition (littoral, deltaic, shelf, oceanic), the biological environment, etc.

FAD (1) (geol.) first appearance datum, see →biohorizon; (2) (biochem.) flavin adenin dinucleotid.

fallout (radioactive) during the above-ground bomb-test nuclear explosions of the years 1945 until 1963, when the nonproliferation act was signed, radioactive fission and neutron-activated products were introduced into the higher atmosphere and redistributed all around the globe. Subsequent precipitation has taken place for decades, reflecting fission isotopes such as ^{137}Cs (cesium) and ^{90}Sr (strontium) as major components and ^{106}Ru (ruthenium), ^{147}Pm (prometium), and ^{144}Ce (cerium) as minor components. ^{239,240}Pu 1.5 10^{16} Bq was produced in total) is one of the transuranics formed during the nuclear explosions, while ^{60}Co (cobalt) is one of the neutron-activated isotopes produced. After 1964, fallout also contained ^{238}Pu (4.9 10^{14} Bq) derived from the burn-up in the atmosphere of the SNAP-9 US space device, using this isotope as energy source. At present the ratio ^{238}Pu/^{239,240}Pu in fallout is 0.04, and any deviation from this value indicates other anthropogenic sources, such as effluents of nuclear reprocessing plants. (Figure see p. 108, 109)

fan see →deep-sea fan.

fathom unit of water depth used on fishing charts, 1.8288 m. At the time when depth was sounded with a weighted line one fathom was the distance between the hands of the sailor stretching both arms (French: bras; English: broad, hence "broad-fourteens", ca. 25 m deep sea area off the Dutch coast).

fatty acids organic aliphatic acids, found in fats and other →lipids, are of various types: (a) straight-chain saturated acids, with the general formula $C_nH_{2n}O_2$ and (b) straight-chain unsaturated acids with one or more double bonds in their molecules. Almost all fatty acids, both straight-chain and cyclic, saturated and unsaturated, found in nature contain an even number of carbon atoms.

fauna (1) name of a rural goddess in Greek mythology, sister of Faunus, used first by Linnaeus in the title of his work *Fauna suecica* (1746), a companion volume to his *Flora suecica* (1745); (2) collective term applied to the animals or animal life of any particular region or epoch; (3) treatise upon the animals of any geographical area or geological period.

fecal pellets the membrane-enveloped feces of marine animals (e.g., →polychaetes, →gastropods, →bivalves, →crustaceans, →teleosts). Large pellets, especially those with →diatom remains, have high →sinking rates, and may reach shelf sea bottoms in 1 to 2 days, forming food for the →benthic fauna (see →coprophagy). Most are of simple ovoid form less than a millimeter long, or more rarely rod-shaped with longitudinal or transverse sculpturing.

fecundity the total amount of eggs produced by a female during a single reproductive cycle. Egg number (N) is usually related to the size (length, L) of individual females, where $N = f L^b$. The power b usually varies around 3 for different species. The fecundity of fish is often estimated in fisheries research in order to calculate the reproductive capacity of a population or, the other way round, to estimate the size of the spawning population from the density of →pelagic eggs and the size of the spawning area.

feeding the act of taking up nutritional substances by an organism, characterized by the speed at which food particles are ingested (feeding rate) or removed from the water (filtering rate) as well as by the origin of the food and the way in which it is collected. The food particles may be of animal origin (→carnivorous feeding), vegetable origin (→herbivorous feeding) or a mixture of both (→omnivorous feeding). The feeding type refers to how the food is gathered; examples are organisms that filter particles from the water (→filter feeding) or ones that seize prey (→raptorial feeding). Another characteristic is the place where the food is collected. Some pelagically living species may collect their food from the sea bottom (→benthic feeding). Feeding is in many marine species, e.g., demersal fish, →zooplankton, restricted to certain hours of the day (→diurnal feeding periodicity), which in the case of zooplankton is accompanied by vertical migration of the species.

feeding rate see →feeding.

Fennoscandian ice sheet see →ice sheet.

fermentation →anaerobic metabolic process mostly in microorganisms, fermenters, in which organic matter is degraded to alcohols, lower fatty acids or ketons. In this fermentation process organic matter is both electron donor and acceptor. The biochemical yield of ATP in fermentation is much lower than during respiratory processes. Well known is the alcoholic fermentation by yeasts in wine and beer production. Examples of alternative fermentation pathways are also found in invertebrates, capable surviving

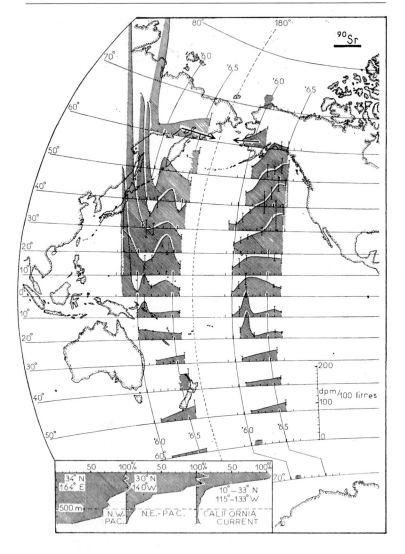

Fallout. Last world overview of radioactive ^{90}Sr fallout concentrations in ocean surface waters. Mean values with standard deviation for 10° latitude zones. For the Pacific seperate eastern and western zones (E and W of 180° longitude). The high peaks of the

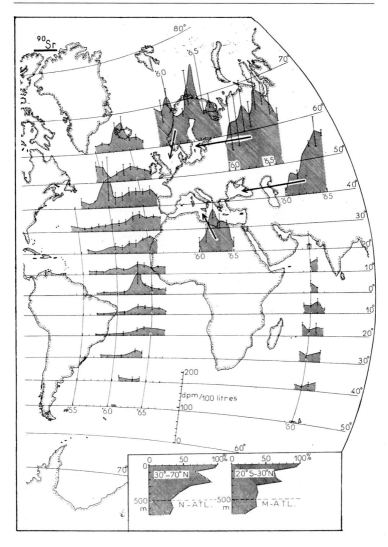

northwest Pacific in the early 60s are direct introductions by nuclear bomb-test explosions; the rest is fallout from the stratosphere. The US, UK, USSR atmospheric nuclear bombtests were stopped in 1963. (Duursma 1972)

anaerobic conditions, e.g., several →Platylhelminthes, →Polychaetes and marine →Bivalvia. These facultative anaerobic invertebrates produce fermentation products such as acetate, ethanol, alanine, alanopine, strombine, lysopine, octopine, or lactate.

fermentor large growth vessel used to culture →microorganisms.

ferromanganese nodules see →manganese nodules.

fertilization the melting of a haploid male gamete (sperm) and a female gamete (egg) into a diploid →zygote or fertilized egg. Fertilization may take place inside the female after sperm has entered during the process of copulation, or outside the female when eggs and sperm are released simultaneously during the process of →spawning. Fertilization usually takes place within a few minutes. In some animals the female can store sperm after one copulation, in order to fertilize several batches of eggs.

fertilizer chemical compound which stimulates growth of plants (nitrate, ammonia, phosphate, urea). Applied in agriculture to increase crops. Unwanted side-effect of fertilizer application is that it can lead to →eutrophication of natural waters.

fetch see →wind waves.

Filibranchiata primitive →Bivalvia, in which the →ctenidia are greatly enlarged for use as a feeding mechanism, and in which the junctions between the gill filaments are either absent or formed by interlocking groups of →cilia. The foot is small and the →byssal glands are well developed. There are one or two adductor muscles. Example: *Mytilus edulis*, the common mussel.

film (surface) see →surface film.

filter feeder (syn. suspension feeder) animal that collects its food by filtering suspended particles from the water using (part of) the organic particles (→phyto-, →zooplankton, →bacteria, →detritus) as food. Many different types of screening apparatus are developed in a wide range of →pelagic and →benthic organisms. The particles are collected from the suspension by hairy appendages, long tentacles, mucus, etc. Each species usually feeds on a specific size range of particles enabling many species to use the same food source (food resource-partitioning). The filtering rate (or clearance rate) is measured as the volume of water filtered (cleared) per unit of time.

filtering rate see →filter feeders.

fin-clipping marking method in studies of fish populations: a small part of the tail fin or another fin of the fish is removed, leaving a distinguishable mark. The fish can be recognized individually since the mark can be made on a particular combination of fin rays. Fin clip marks will disappear, usually within 1 or 2 years, because the fish have the capability to heal.

first appearance datum see →biohorizon.

fish see →Pisces.

fish culture the production of fish in an artificical surrounding controlled by man. Fish can be cultured in: (a) tanks; (b) ponds with either stagnant or running water, varying with fish species and locality. Pond cultures can yield 3800 to 5000 kg ha^{-1} annually); (c) cage-cultures, installed in stagnant water lakes, in lotic systems (creeks and rivers) or in marine tidal areas (fjords). The production depends on the fish species and type of cage culture, but may amount to 350 kg m^{-3} annually, as in the case of carp; (d) large enclosures, often divided into compartments by nets. In these culture systems, productions of up to 4000 kg per ha can be reached. The culture set-ups may be stocked with larvae or young fish collected in the natural system, or with brood produced in special hatcheries where fertilized eggs are incubated and larvae reared to suitable size for stocking in production units. In extensive cultures, fish production depends on the natural food production of the system, as in the culture of most →herbivorous fish (milkfish, mullets). In intensive cultures fish production depends on the additional supply of food for mostly →carnivorous species (salmon, eel, trout). Factors that raise →growth rates in cultured fish above natural levels are optimal temperature and excess food of optimal composition, but it is most of all the decreased →mortality in the controlled environment that leads to the high fish production required to make fish culture worthwhile. Sometimes fish culture concentrates on rare and expensive species. See also →aquaculture.

fish farming commercial fish culture on a large scale. The farming can include the production of brood in special hatcheries. See also →fish culture.

fish production the →biomass produced by a fish →population or →community per unit of space and time. In →fishery science or ecological research the term is used to indicate total production which besides somatic growth includes the generation of reproductive materials, and also the production lost through →mortality.

fish trap construction of fences (steel or wood) across rivers and streams, leading migrating fish into a fyke or small chamber where they can be collected. (Figure see p. 112)

fisheries the human activities aimed at catching fish. There are many types of fisheries, depending on the fish species, the developmental stage of the human society, and the hydrography and geography of the sea area. A first distinction is that into fisheries using passive and active catching methods. Passive are baited angles, long lines, fish traps (as, e.g., lobster pots), but also drifting nets and fykes, made of netting or wood, and →fish traps in rivers. Active methods range from the simple hunt with arrows and spears,

Fish trap.
(Müller 1982)

the enclosing of shallow water with nets, to the fishing with modern trawlers operating beam trawls, →pelagic trawls, danish seines and purse seines. One of the most peculiar ways of fishing is with trained cormorants wearing neck-rings to prevent them from swallowing the fish they bring to the small boats of their owners (China, Japan).

fisheries convention convention on fishing and conservation of living resources of the high seas, signed by 58 countries in Geneva, 1958, and coming into force in 1966 after the required minimum ratification by 22 countries. The convention considers the development of modern techniques for the exploitation of living resources of the sea, increasing man's ability to meet the need of the world's expanding human population for food and the exposure of these resources to the danger of being over-exploited. Other bilateral and multilateral conventions on fishery rights outside the 12 miles' →territorial sea exist.

fishery science fisheries research, field of research dealing with (a) the dynamics of commercially exploited →populations of fish and invertebrates (i.e., fisheries biology); (b) biological oceanography as far as relevant to a, and (c) fishing techniques and materials and their possible improvement. The principle aim of fisheries research is to determine the present effects of fishery on the fish populations and, to indicate what management of the fisheries would lead to maximum sustainable yields. The work of fisheries laboratories, therefore, concentrates on the quantification of a number of properties of commercially important fish populations and their exploitation. The size and age composition (see →age determination) are estimated from the commercial catches (sound →statistics of which are essential) and other sources like young fish surveys, →acoustic surveys and →tagging experiments. The production of fish populations is estimated on the basis of many different kinds of information, as e.g., the fertility, egg production, survival of the early life stages, →recruitment of young fish to the exploited stock, →growth rate as a function of age and size, and on different kinds of

→mortality. The latter include natural mortality (estimation of which requires study of the population size, →gut content and →digestion rate of all possible predators) and fishing mortality (estimated from the catches and the discards). Mathematical →models are an important tool of fisheries scientists; at present the conventional population dynamical models are replaced by multispecies models in order to cope with the complexity of the whole of mutually interacting fish populations under exploitation. Because most fish resources in the open sea are exploited simultaneously by different countries, and since single fish populations can inhabit or, by →migration temporarily inhabit, different national waters, governamental fisheries research activities have since the beginning of this century been coordinated by the International Council for the Exploration of the Sea (ICES), which has its headquarters in Copenhagen.

fjord long narrow inlet or sea-arm, steep-walled and with a U-shaped profile due to previous glacier erosion often several hundred meters deep. A fjord is typically situated in a mountainous coast at higher latitudes. It has a shallow sill or threshold near its mouth and becomes deeper inland. Due to this morphology the deep water layers may be anoxic. See also →coast.

Flagellata flagellates, see →Mastigophora.

flagellum (pl. flagella) a long hair-like lash, which by a rowing or undulating motion draws or propels the body or attracts particles; responsible for movement in unicellular organisms (→flagellates) and used for water movement through the body in →sponges. See also →cilium.

flame cell hollow organ with a bundle of →cilia which work in the lumen. They are usually connected by canals which ultimately open to the exterior.

floating adaptations see →buoyancy.

flocculation the formation of →aggregates as a result of either the decrease of the repulsive forces between electrically charged particles (clay particles and hydroxides), or microbiological action on organic matter. The decrease of charge is caused by a change in the electrical double layer. Flocculation can be stimulated by an increase in the salinity of water. It is a reversible process; the decomposition of the aggregates is called deflocculation.

flood tide the period in a tidal cycle between low water and the next high water. Flood current is the current during the rising of the sea level. In the case of a →progressive tidal →wave the reversals of the current take place just between the times of high and low water, and a flood current cannot well be defined.

flora (1) the goddess of flowers in Greek mythology; (2) a descriptive catalog of the plants of any geographical area or geological period; (3) the plants or plant life of any particular region or epoch.

flow (1) in general current, e.g., of water, but also of energy, carbon in a system, etc; (2) mass movement of unconsolidated material that exhibits a continuity of motion and a plastic or semifluid behavior resembling that of a viscous fluid. Water is required for most types of flow movement. Also, the mass of material moved by a flow. See →mudflow and →sediment-gravity flow.

flow separation in a quasi-stationary fluid flow along a boundary, it may occur that at a certain point the →streamline following that boundary loses its contact due to a local curvature or edge of the boundary surface. This flow separation is caused by the inertia of the fluid particles. Flow separation in air flowing over →white-capping waves is thought to be an important mechanism for momentum transport from the air to the sea.

fluidized sediment flow see →sediment-gravity flow.

fluorescence microscopy a method using the fluorescence of objects in microscopic examination. While some objects can be autofluorescing, e.g., chlorophyll-bearing bodies such as chloroplasts, others must first be stained with fluorochromes such as acridine orange. Light of specific wave lengths is required to generate fluorescence, and in microscopy the quality of light is controlled through a set of filters (exciter filter, beam splitter, barrier filter). Fluorescence microscopy is used in biological oceanography to count, e.g., →Bacteria, →Protozoa and →phytoplankton.

fluorescence of seawater irradiating seawater with ultraviolet (UV) light results in a weak emission of fluorescent light in the visible part of the spectrum. This fluorescence was first determined by K. Kalle in the early 1950s, using a simple Pulfrich photometer. Fluorescence is expressed in units of mFl, where 1 mFl corresponds to the fluorescence of a $1.43 \, 10^{-3}$ mg chinine-bisulfate dm^{-3}. In fact, →humic or →fulvic compounds are responsible for this fluorescence, and by using this method of detection, these compounds can be followed when discharged from terrestrial sources into the sea. Part of the fluorescent substances in the oceans are also →autochthonic. See also →phosphorescence.

food chain at the start or basis of the food chain, organic matter is produced by plants (algae in the sea). It is eaten either alive by →herbivores or dead by →detritivores. These consumers of vegetable material are in turn eaten by →carnivores which are then eaten by top carnivores (or top predators), being the end or top of the food chain. In such a way, organic matter is transferred along the links (called →trophic levels) of a food chain. This transfer is far from 100% efficient, because living organisms need most of their food for maintenance (see →production efficiency). In practice, food chains are rarely linear, but are branched and interwoven. Therefore, a more proper name is food web.

food web see →food chain.

Foraminifera [foraminifer(an)s, forams] order of the subphylum →Sarcodina (phylum Sarcomastigophora, see →Protozoa), unicellular marine

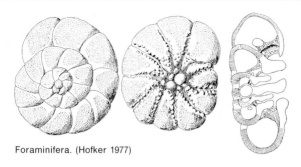

Foraminifera. (Hofker 1977)

organisms living either on the bottom (→benthic) or floating in the water column (planktonic). They have either a usually multi-chambered shell, or reticulate pseudopodia, or, usually, both. Planktonic species have a vacuolated outer layer of protoplasm. The shells are calcareous, but are occasionally chitinous, siliceous, or gelatinous with embedded foreign bodies. Planktonic foraminifera, e.g., *Globigerina* are so numerous that their empty shells form the →ooze which covers enormous areas of the ocean floor, particularly in the Atlantic, Indian, and Southern Oceans. They have existed since the →Cambrian and are widely used by geologists for →biostratigraphical zonation. Striking is the extinction of the Globotruncanidae at the →Cretaceous-Tertiary boundary. See →biostratigraphy and →microfossil.

forced wave a wave that can only exist because of the permanent local presence of an external forcing. This can be compared with a pendulum moving under the influence of an oscillating force with a frequency that differs from the pendulum frequency. If the external force stops, the forced wave will disappear and only →free waves are left. Waves that only exist because of the oscillating boundary conditions like the tidal wave in a marginal sea such as the North Sea are considered to be free waves, since the forcing is not locally present. This in contrast to, e.g., the Southern Ocean, where the tidal forces locally generate a forced tidal wave. Hereby energy is transferred from the forcing system to the wave motion.

fouling growth of sessile algae and animals, especially on a ship's bottom and other man-made underwater structures.

f-plane approximation see →Coriolis parameter.

fracture zone zone of →transform faults in the ocean, floor offsetting the axis in the ocean. See also →plate tectonics.

free wave a wave that has been generated outside the area of occurrence or at an earlier time, and that, time and area considered, is independent of the presence of a local external forcing: this in contrast to a →forced wave. Energy input into the wave only occurs by an energy flux across the boundary of the definition area.

free-air anomaly see →gravity anomaly.

freshwater natural water, derived from rain or melted snow, with low concentration of dissolved minerals. The chemical composition may be highly variable, in contrast to seawater, in which the concentration ratios of the major solutes are stable.

Freundlich and Langmuir isotherms for sorption-desorption processes, partition can be described, otherwise than by →distribution coefficients, by either the Freundlich or Langmuir isotherm. They are, however, only applicable in systems of reduced complexity, where the Freundlich isotherm has an empirical basis and the Langmuir isotherm is considered to have a thermo-dynamic background. The Freundlich isotherm is given by $C_{ads} = K\,C_{sol}^{n}$, K and n being constants; the Langmuir relation is given by $C_{ads} = K_1\,C_{sol}/(1+K_2 C_{sol})$, K_1 and K_2 also being constants. At dilute conditions both relations allow for an almost linear representation when C_{ads} is plotted against C_{sol}, thus approaching proportionality between these concentrations with the →distribution coefficient K_d as a constant factor.

friction is the process by which large-scale motions are transformed into heat. In the case of a current flowing over a bottom friction is caused by the turbulent transport of momentum towards the bottom. This gives a bottom stress opposed to the current direction which depends quadratically on the current speed. The roughness of the bottom is another important factor in bottom friction, and its effect is represented by a so-called drag coefficient. For some problems, somewhat artificially, a linear relation between current speed and friction is assumed (Rayleigh friction). In the interior of a fluid, friction also may occur as the result of current →shear. The transport of momentum from one layer to the other by →turbulence depends on the shear and the strength of the turbulence, expressed by the →eddy viscosity.

fringing reef see →reef.

front a front or a frontal zone is a sloping interface between two water bodies with different properties. Fronts may persist for some time when the →advection at either side of the front counteracts cross-frontal mixing. This means that across a front the current component at right angles will change rapidly. There are different types of fronts. The larger oceanic frontal zones are the result of large-scale advection of different water masses, e.g., between the subtropical and the subpolar →gyre. Coastal fronts develop when water of estuarine origin flows into the open sea. Thermal fronts may develop in shallow water between vertically well-mixed water and stratified water further offshore. The converging water at the surface is effective in concentrating foam or flotsam at the surface along the front. Concentration of particulate matter along a front may also increase biological activity.

Froude number a nondimensional number occurring in →similarity theory and used in model experiments in open channels, where the hydrostatic pressure plays a role. The Froude number (named after the British engineer W. Froude, who introduced scale experiments in ship building in 1871), Fr, gives the ratio between the current speed u and the speed of propagation of a →long wave, c, in water of a depth D, and equals $Fr = u/c = u/\sqrt{(gD)}$. It also gives a criterion between two possible types of flow, subcritical (the normal condition in oceanography) and supercritical flow, with Fr respectively smaller and larger than 1. A tidal →bore is an example of the transition between the two types of flow. The Froude number is equal to the inverse square root of the bulk →Richardson number, applied to the total layer of water.

frozen-field hypothesis hypothesis in →turbulence studies, implying that the time variations of the turbulent field at a given point are similar to the spatial variations at each moment. So similar time variations would be found if the turbulent field were suddenly "frozen" and moved along this point with the mean velocity.

frustule see →Bacillariophyceae.

fucoxanthin see →carotenoids.

fulvic acids these ill-defined naturally occurring acids are part of organic matter (humus) (fulvus = yellow). The compounds contain →hydrocarbons (n-alkanes) with a carbon range of C_{14} to C_{31}, which is also characteristic of →fatty acids in marine organisms. Fulvic acids can be arbitrarily distinguished from →humic acids by fluorescence spectrometry. See also →humic acids.

Fungi molds, yeasts, and mushrooms forming one of the five kingdoms of the living organisms. They are →heterotrophic, →eukaryotic, filamentous, or multicellular, and differ from plants by the absence of →chlorophyll and from bacteria by the fact that fungal cells are much larger and possess vacuoles, nuclei, and other intra-cellular organelles typical for →eukaryotic cells, which can be seen easily by the ordinary light microscope. The microbial community in marine sediments can at certain times be dominated by fungi, especially in marine waters subjected to freshwater input containing fungal spores.

G

Gaia name of the Greek goddess of Earth. For Gaia Hypothesis see →holistic approach.

Gammaridea (gammarids) see →Amphipoda.

ganglion small mass of nervous tissue containing numerous neurons (i.e., nerve cell bodies). The nervous system of many invertebrates consists largely of such ganglia, connected by nerve cords, variously arranged but usually well-developed in the head (cerebral ganglia). In vertebrates ganglia also occur in the peripheral and autonomous nervous system.

gas bladder (swim - or air bladder, air sac), organ in fish of which the original function was probably a respiratory one evolved into a hydrostatic organ. By adjusting the gas content (air or gas secreted in its lining) specific gravity of the animal is changed, and with that the →buoyancy.

gas chromatography an analytical technique based on the partitioning of a solute into a mobile gas phase and a stationary liquid phase. The stationary phase is coated on the inner wall of a column or on the surface of a carrier. Solute sample and mobile phase are introduced into the entrance of the column. The exit of the column is connected to a detection system. Partitioning of the solute between the mobile and the stationary phase is determined by the temperature, the chemical/physical properties of the solute and the liquid phase. Volatile compounds with low affinity for the stationary phase will leave the column first. The technique is commonly used for the separation of complex mixtures, such as organochlorines, into their constituents.

gas vesicles occur in a number of →prokaryotic organisms that live floating in lakes or the sea and which produce gas vesicles which make floating or →buoyancy possible. Floating due to gas vesicles is sometimes seen in massive accumulations (→blooms), where gas-vesiculate cells rise to the surface and can be driven by the wind into dense masses.

gases, dissolved a useful assumption in chemical oceanography is that at some time any parcel of water in the world's ocean has been at the sea surface, and become equilibrated with atmospheric gases. Therefore equilibrium concentrations are defined at atmospheric pressure and not at →in situ pressure. According to Henry's Law, which holds for nonreactive gases, the equilibrium concentration (c) at a certain temperature in a liquid is proportional to its partial pressure (p) in the gas phase: $c = \alpha p$. The solubility coefficient α is characteristic of the gas; it decreases with increasing temperature and with increasing salinity. The distribution of the concentration of gases in the oceans are governed by the global (meridional) water circulation, by the meridional temperature differences at the surface and, for gases like →oxygen, by →in situ processes which consume or produce the gas involved.

gases, in air the total pressure of the atmospheric gases at sea level approaches 10^5N m^{-2}(1 atm). The components (except for water vapor) in volume percentages are for N$_2$: 78.084$\pm$0.004; O$_2$: 20.946$\pm$0.002; Ar: 0.934$\pm$0.001; CO$_2$: 0.033$\pm$0.001 and expressed in ppm volume for: neon: 18.18$\pm$0.04; helium: 5.24$\pm$0.004; krypton: 1.14$\pm$0.01; xenon: 0.087$\pm$0.001; H$_2$: 0.5; CH$_4$: 2; N$_2$O: 0.5$\pm$0.1.

gastric digestion the process of food digestion in stomach and intestines (gastric system) of animals.

Gastropoda (gastropods) class of the phylum →Mollusca. Snails. The shell, unlike that of the →Bivalvia, is an asymmetrical spiral, the axis of which is the columella; the shell may be lost secondarily. When crawling on the extended foot and feeding, the gastropod carries the shell; when disturbed, the body is completely withdrawn into the shell, which is closed with the →operculum. Gastropods live on land, in freshwater, and in the sea, →benthic as well as →pelagic, and are the most successfully developed group of mollusks. They are divided according to the type of gills into the subclasses →Prosobranchia (two gills), →Ophistobranchia (one gill), and Pulmonata (gills absent, lung developed from →mantle), with marine snails among all types. The abundance of gastropods as fossils is enormous; but there are even more recent species known (ca. 35000).

Gastrotricha phylum of microscopic →Aschelminthes, in the marine environment living in the interstitial water in bottom sediments (see

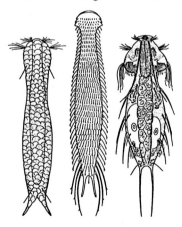

Gastrotricha. (Barnes 1987)

→meiofauna). Unsegmented worms with a flattened body, with →cilia on the ventral side of the trunk and head for locomotion. →Hermaphrodite or →parthenogenetic. May be confused with →Ciliophora, and superficially also resemble →Rotifera.

Gauss epoch, - chron see →magnetostratigraphy.

gelatin water-soluble protein, yielded by →hydrolysis of collagen (protein of connective tissue in animals).

Gelbstoff (syn. yellow substance) term introduced by K. Kalle (see →fluorescence of seawater), who noted the absorption of light in the blue part of the absorption spectrum of both sea- and freshwater containing much dissolved organic matter (see →organic carbon), resulting in yellow water. Gelbstoff is often used as a synonym of persistent substances, but this has neither a quantitative, nor a qualitative value. Related to →humic acids and mainly transported to sea by land run-off. Gelbstoff can be used as a tracer of river water in →estuaries and the sea.

generation time see →population dynamics.

geobiocoenosis see →biogeocoenosis.

geochemical cycle for a given chemical element or compound its distribution and transport over the surface of the earth can be described in a mass balance approach. Typical reservoirs are the continental →crust, continental sedimentary rocks, freshwaters, seawater, the atmosphere, the continental and marine biosphere, deep-sea sediments and Mid-Ocean Ridge Basalts, etc. Fluxes (mol y^{-1}, Gtons y^{-1}) between reservoirs are quantified in rate terms of weathering, river transport, estuarine removal, seawater input and removal, evaporation, precipitation, sedimentation, submarine hydrothermal weathering, etc. For a given reservoir the ratio of its size and its overall throughput (fluxes) yields the characteristic residence time. The →residence time of a chemical element in the ocean may vary from about a decade to millions of years.

geochronology measurement of geologic time, either in →absolute age or in relative age (see →geological time scale and →stratigraphic hierarchy).

geoid (syn. geopotential surface) the theoretical (equipotential) surface of the earth which is everywhere perpendicular to the direction of →gravity (the plumb line). It can be considered a mean sea-level surface extended continuously through the continents. If the oceans were everywhere at rest the ocean surface would be a geopotential surface.

geological time scale scale of relative time. The division is based on biostratigraphic and lithostratigraphic characteristics. Names used here can

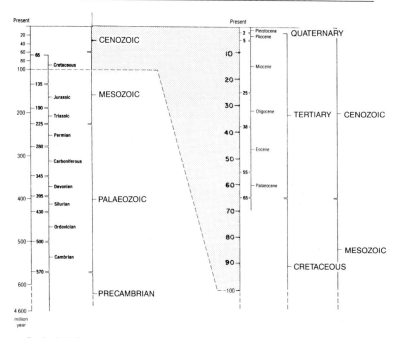

Geological time scale. (After Open University 1989)

be used as a unit of time and also to designate the strata formed in that unit (see →stratigraphic hierarchy).

geopotential surface see →geoid.

GEOSECS (Geological Ocean Sections) acronym for an international suite of expeditions in the late 1960s, early 1970s, largely through the US National Science Foundation, with the purpose of mapping the distributions of dissolved tracers in the worlds' oceans.

geostrophic balance if the →Coriolis force on a moving fluid is fully compensated by the pressure force, the fluid is in geostrophic balance. In that case the current is parallel to the →isobars, clockwise around high pressure centres in the northern hemisphere and anticlockwise in the southern hemisphere. Buys-Ballot's (1817–1890) law in meteorology is the consequence of geostrophic balance. In the open ocean the assumption of geostrophic balance is often applied to the pressure field derived from the temperature and salinity observations to calculate the current relative to the →level of no-motion. Exact geostrophic motion would result in stationary flow. In reality apparently geostrophic flow changes gradually, so there is a small deviation from the balance. These conditions are called quasi-geostrophic motion.

geothermal heat flow heat flow from the inner earth. It can usually be neglected in heat calculations, compared with the other fluxes, but it is used in geophysical and geological research as a way to obtain information of the subbottom structure. Its average over the whole earth is 1.2 10^{-6} cal cm^{-2} s^{-1}.

Gilbert epoch, - chron see →magnetostratigraphy.

gill nets see →nets.

glacial epoch (syn. ice age) any part of geologic time in which the climate was significantly colder than today. Glacial epochs are characterized by widespread glaciers (→ice shelves and →ice sheets) which moved from the poles towards the equator and covered a much larger area than today. The maximum of the most recent glacial epoch was 18 000 years ago.

glaciation the formation and advancement of →glaciers and →ice sheets towards lower altitudes, and action on the earth's surface by glacial activity.

glacier large mass of ice formed, at least in part, on land by compaction and recrystallization of snow and surviving from year to year. Glaciers are able to flow by creep under the influence of gravity. The term includes not only the small mountain glaciers but also →ice sheets, →ice caps, and →ice shelves.

glaciomarine sediments sediments transported in or on glaciers or floating ice and deposited at the sea floor. Sediments transported in this way are often very poorly sorted in →grain size. Because of their particular character they are easily distinguished from other marine sediments.

glauconite →clay mineral, a green →illite that can be →authigenically formed in marine sediments. Because both Fe^{2+} and Fe^{3+} occur simultaneously, the mineral is formed in microenvironments which are alternately reducing and oxidizing, e.g., within the tests of →Foraminifera.

globigerina ooze see →Foraminifera, and →ooze.

gonad organ producing gametes (reproductive cells: eggs and sperms).

Gondwana the late →Mesozoic continent that split into present-day South America, Africa, Australia, India, and Antarctica. See also →continental displacement and →Pangea.

gonoduct tube through which the mature germ cells pass to the exterior.

grab ocean-bottom sampler that consists of segments which drive into the sediment, enclose and retain a sample. The Van Veen grab is made up of two opposed, pivoting, hemicylindrical buckets that shut by rotation on a hinge when the sampler is pulled up. A fair-sized grab can bring up approximately 1 m^3 of material. See also →bottom samplers.

graded bedding beds of sediment in which relatively coarse material is concentrated at the base and finer material at the top. They may form when the velocity of the prevailing current oscillates (each level will then be well sorted), or by emplacement by a single short-lived flow with declining velocity, such as a →turbidity current.

grain flow see →sediment-gravity flow.

grain size size of sediment particles; constitutes the main characteristic used for classifying lithogenous sediments. One can distinguish: →clay – smaller than 2 μm; silt – between 2 and 64 μm; →sand – between 64 μm and 2 mm; granules – between 2 and 4 mm; gravel pebbles – between 4 mm and 6 cm; cobbles – between 6 and 25 cm; and boulders – larger than 25 cm. See also →grain-size distribution.

grain-size distribution percentage (by weight or count) of particles of each →grain-size fraction. The grain-size distribution is a measure of the energy of the depositing medium and the "energy" of the basin of deposition. In general, coarse sediments are found in high-energy environments and finer sediments in low-energy environments. The availability of grains of various sizes largely determines the validity of the measure.

gravel see →grain size.

gravity the resultant force on any body of matter due to both the attraction by the earth and the centrifugal force resulting from the earth's rotation about its axis. Gravity is defined by the acceleration of a freely falling body. It is measured by gravimeters (measuring the stretching of a spring). At sea such measurements were impossible for a long time, because of the motions of the ship by waves. The first reliable gravity measurements at sea were made by F.A. Vening Meinesz (1887–1966) with a compensated pendulum apparatus from a submerged submarine. He discovered the belt of negative →gravity anomalies accompanying →island arcs. Presently gravimeters can be operated on surface ships by using servo-mechanisms that keep the instrument well-oriented and by strongly damping the oscillations of the instrument. See also →geoid.

gravity anomaly any discrepancy between the measured terrestrial gravity and the theoretical gravity at some reference surface, usually the so-called reference ellipsoid. The theoretical gravity and position of the reference ellipsoid is calculated as if all masses in the earth's interior were perfectly regularly distributed, according to a fluid in rotation. Excess observed gravity is called positive and denotes an surplus of mass, whereas a negative anomaly denotes a mass deficiency. Free-air anomaly is the anomaly resulting if the measured gravity is corrected for the elevation of the earth's surface with respect to some level datum (not necessarily the reference ellipsoid). The Bouguer anomaly is the anomaly resulting after corrections for elevation and the attraction of the rock mass between station and level datum.

gravity corer see →corer.

gravity wave in a stratified but nonrotating fluid the only restoring force in case of a deviation from dynamic equilibrium is the →buoyancy force. It causes oscillations within the water around an equilibrium state that are called →internal gravity →waves. At the surface the gravity is the restoring force for wavelengths larger than some centimeters. For smaller wavelengths gravity waves are replaced by →capillary waves.

gray (Gy) unit of absorbed radiation dose, used to specify the radiation absorbed by the human body or parts of it, equal to 1 J kg^{-1}. Replaces the old rad (1 Gy = 100 rad).

grazing in the marine environment, the ingestion of →phytoplankton or sedentary →algae by marine animals. These →grazers filter the water (→sponges, crustaceans, mollusks, salps) or rasp the substrate (→sea urchins, gastropods). The amount of grazing is experimentally measured by the decline of food in the ambient medium (→chlorophyll, or number of algal cells via →particle counters) or by the increase in the animal of ^{14}C from labeled food, or of degraded chlorophyll in the →gut content). Overgrazing (ingestion larger than the digestion can cope with) may occur when food (phytoplankton) is very abundant, resulting in →fecal pellets rich in undigested algal cells.

Great Barrier Reef see →reef.

green algae see →Chlorophyceae.

greenhouse effect the elevation of the temperature of the earth and its atmosphere due to the presence in the atmosphere of so-called greenhouse gasses (CO_2, CH_4, N_2O, CFC's like the Freons, O_3) which diminish the reradiation of heat into space. The expected rise in global temperature, due to increased concentration of these gases, ranges from 1.5 to 4.5 °C within the next 100 years and may be one of the major long-term environmental problems. In the marine environment this is expected to cause a rise in sea level of 0.5 to several meters, due to the expansion of seawater at higher temperature, melting of snow and ice on land, and possibly, but not yet likely, breaking-off of quantities of Antarctic ice, which will melt at lower latitudes. The difficulty of measuring this temperature increase, due to the large interannual variations (→climatic variation), makes the greenhouse effect difficult to prove, hence the uncertainty in the prediction of its size and effects. Changes of the temperature of seawater will also greatly affect areal distribution of marine organisms.

gross production see →production, primary.

group velocity the velocity at which →wave energy is propagating. For →short waves the group velocity is half the phase velocity (see →wave characteristics), for →long waves it is equal to the phase velocity. For pure

→capillary waves the group velocity is one and a half times the phase velocity.

growler see →iceberg.

growth increase in size, used both for individuals and populations. Individual organisms may grow at equal rates in all directions (isometric growth) or growth may be allometric, i.e., stronger in certain directions such as length. Populations grow in size by reproduction (→natality) and sometimes also by →immigration. At unlimited food supply, initial small numbers increase at first at an accelerating rate until food (or other resources) becomes limiting and the rate of increase becomes lower and lower. This results in S-shaped →growth curves.

growth curve graph showing length (or any other one-dimensional measure of body size) or weight (→biomass) of the average individual organism plotted on a time axis. The simplest way to construct a growth curve is to fit a curve through a series of frequently measured lengths or weights. Study of the basic concepts of the process of growth has shown that individual growth can be explained as a result of size-dependent anabolic and catabolic (see →metabolism) processes that compensate each other at maximum size W^n, and mathematical equations have been developed to describe the process. A well-known equation is that developed by Pütter, Bertalanffy and Taylor: $w = [W^m - (W^m - W_0^n)e^{-nk(t-t_0)}]^{1/n}$ which has the sigmoid form (S-shaped), and is used to calculate production for a great variety of animals, especially in →fisheries research. See also →growth.

growth efficiency ratio between growth (measured as weight increment) and food intake (measured in the same units, e.g., grams ash-free dry weight). Animals grow efficiently if they need little food for fast growth. Young animals generally show higher growth efficiency than old individuals of the same species. Compare →production efficiency for a similar ratio for entire →trophic levels of an →ecosystem.

growth form (1) used for the growth of populations, two basic patterns can be discerned based on the shapes of arithmetic plots of growth curves: the J-shaped and the S-shaped or sigmoid growth form, see →growth and →growth curve; (2) used in ecological research of, e.g., reef corals where colony form is adjusted to the environment, particularly to water movement. Massive, encrusting, and stoutly ramose or branching growth forms occur in exposed habitats, delicately branching or thinly foliaceous forms in sheltered habitats. Study of growth forms in recent and fossil faunas often gives a better insight into the (→paleo)environment than species lists. In →Bryozoa some 20 growth forms can been discerned.

growth rate in organisms the average increase of weight per unit of time. In →populations the number of organisms added to the population per unit of time. The specific growth rate of a population is the growth rate divided by the number of organisms initially present or by the average number of organisms during the period of time.

guano dried sea-bird excreta (a mixture of urine and feces), piled up at undisturbed and densely packed seabird colonies on arid, subtropical islands. Guano islands are typically situated in highly productive upwelling areas which support massive food-stocks and huge colonies of seabirds. Guano-producing seabirds often withhold their excreta for nest building. A Guanay cormorant (*Phalacrocorax bougainvillii*) deposits a minimum of 1 kg dry guano on land each month. From the 19th century onward, guano has been commercially exploited as an inexpensive fertilizer that is rich in nitrogen, phosphate and calcium. Whole seabird colonies have been destroyed through reckless exploitation of the resource (including the birds themselves). Between 1848 and 1875 more than 20 million tons of guano were exported to Europe and the USA from Peruvian deposits. Today, guano is harvested as a renewable resource, that accumulates at about 8 cm per year. The main birds involved are cormorants, gannets and boobies, pelicans, and penguins.

Gulf Stream strong →boundary current along the east coast of N. America, carrying water from the Gulf of Mexico northeastward. The name Gulf Stream is usually reserved for the part of the current that has the typical character of a strong boundary current and the continuation to the east of New Foundland is better indicated by North Atlantic Current, which splits into different branches. Part of it is returned and forms the North Atlantic →gyre, part of it feeds the Norwegian and other arctic seas.

gut content the amount of food present in the alimentary tract, expressed in terms of weight (mostly ash-free dry weight), or numbers of different food items. A difficulty is the variable degree of digestion; sometimes different digestive stages are defined for separate determination of numbers or weights.

guyot see →tablemount.

gyre a closed current system of ocean-wide dimensions. A typical subtropical ocean gyre consists of a strong →boundary current at the western branch that carries the water poleward, and a less well-defined return flow in the east. They are driven by the wind system and their asymmetric form is the consequence of the balance of →vorticity.

H

habitat the characteristic space occupied by an individual, a population, or a species. Habitats in the sea are diverse and include parts of such environments as estuaries, tidal flats, intertidal rocky shores, oceanic sea surfaces, deep-sea bottoms, etc. Within such parts of the sea most species occupy only specific zones, e.g., only sandy tidal flats below mean tide level in the tropics.

hadal see →zonation.

haemocyanin see →hemocyanin.

haemoglobin see →hemoglobin.

half saturation constant see →Michaelis-Menten kinetics.

half-life time in which the amount of a →radionuclide decays to half its initial value, a constant for a particular nuclide. See also →radioisotopes.

halocline layer in which the salinity changes rapidly with depth. See also →stratification.

halophilic (halophile) preferring or tolerating high salinities.

halophyte plant species that can tolerate a high salt concentration in its environment. Generally said of plants growing close to the seashore or in →salt marshes.

Hardy plankton recorder see →continuous plankton recorder.

harmonic analysis of tides a method for splitting up the actual tide in its →partial tides. As the periods of these components are known from astronomical considerations, the tide for a given location can be considered to consist of a combination of these components, be it with unknown amplitude and unknown phase with respect to the tide-generating potential. For a registration of sufficient length one may determine amplitude and phase by means of mathematical procedures. For many ports these phases and amplitudes have been established, and subsequently they can be used for tidal prediction at any moment in the future (provided that conditions are unchanged). From these analyses, considered over a larger area, it appears that the dynamic response of a sea area is more important than the effect of the local tidal potential. In most areas the tides are of a →semidiurnal character, although in some seas →diurnal tides predominate. When in a monthly alternation diurnal and semidiurnal tides occur, one speaks of mixed tides. Yet in semidiurnal tides the diurnal components cause a diurnal inequality. In shallow seas also so-called →shallow water tides occur, nonlinearly generated harmonics of the principal tidal components.

heat balance in the oceans heat is directly gained from solar radiation and directly lost by long-wave (thermal) radiation emitted from the surface. In addition, the balance contains the heat exchanged with the atmosphere (→ocean atmosphere exchange), by dissipation of kinetic energy and →geothermal heat flow. Only the radiation and the atmospheric exchange are of practical significance. Time variations in the balance cause diurnal or seasonal warming or cooling restricted to the surface layers. In a long-time average the gain and loss should be equal, if the possibility of a →climatic variation is disregarded. There is a net gain of heat at lower latitudes and a net loss at higher latitudes, with the parallel of 40°N and S as the approximate boundaries. The heat transport of the ocean needed to preserve a zonal balance consists of poleward flow of relatively warm water and an equatorward flow of cold water by the large-scale →circulation in the ocean.

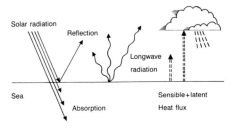

Heat balance

heat capacity see →specific heat.

heat flow see →geothermal heat flow.

heavy metals popular yet incorrect description of several transition metals, with a connotation towards those metals that have been introduced by mankind into coastal and oceanic waters, e.g., cadmium, copper, zinc, lead, etc.

Heincke's law empirical relation, formulated in 1913 by the German marine biologist Heincke (1852–1929), relating the distribution of a fish species with water depth. According to Heincke's law the size of the individuals of a certain fish species increases with increasing water depth and distance from the coast. Based on studies of the Dutch marine biologist Redeke (1873–1945) published in 1904.

Heliozoa order of the subphylum →Sarcodina (phylum Sarcomastigophora), spherical →Protozoa of primarily freshwater. Characteristic are the needle-like pseudopodia, radiating from the cell. Sometimes with skeleton of siliceous material.

Hemichordata phylum of worm-like animals that together with the other small phylum →Chaetognatha, the chordate subphyla →Urochordata and →Cephalochordata form the transition area between invertebrates and vertebrates. Worm-like marine animals subdivided into Enteropneusta (acorn worms) and Pterobranchia. Enteropneusta have cylindrical bodies composed of an anterior conical →proboscis, a collar and a long tail; they live in burrows in the sediment, e.g., *Saccoglossus*. The Pterobranchia contain few species, which are mostly tubicolous and bear a pair of tentaculate arms on the collar that are used in →filter feeding.

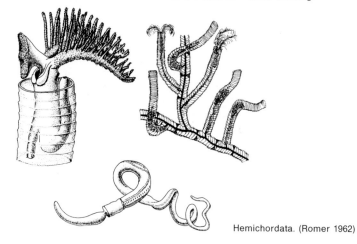

Hemichordata. (Romer 1962)

hemipelagic sediments see →deep-sea sediments.

hemocoel body cavity, actually an expanded part of the blood system, containing blood. Well developed in →Arthropoda and →Mollusca, where the →coelom is small.

hemocyanin (haemocyanin) large, colorless molecules occurring in esp. →Crustacea and →Mollusca as oxygen-binding protein. In hemocyanin, the copper is bound to the protein and does not occur in a functional group; it always occurs dissolved in the plasma.

hemoglobin (haemoglobin), the red coloring matter of blood. The pigment consists of an iron-porphyrin (heme, haeme) coupled to a protein (globin). Hemoglobins consist of variable numbers of unit molecules, each unit containing one heme and its associated protein. The unit molecular weight for vertebrates is ca. $16 \cdot 10^3$; the hemoglobins of higher vertebrates have four units and are intracellular (red blood corpuscles or erythrocytes). Heme is a protoporphyrin consisting of four pyrroles with an iron atom in the center. Hemoglobin is generally associated with oxygen binding, but also possesses an exceptional capacity to react with a variety of other substances, such as organic and inorganic ions (salts) and dissolved gases. The function of hemoglobin depends on the reversible combination of the ferrous (Fe^{2+}) iron with oxygen (oxygenation) in proportion to the partial pressure of oxygen. O_2 combines with each iron atom. The iron-protoporphyrin heme coupled with specific proteins is the best-known oxygen-carrying pigment. The protein part or globin varies considerably in size, →amino acid composition, charge, solubility, and other physical properties from animal to animal. Hemoglobins may be extracellular in body fluids, or intracellular in special corpuscles, or occur in tissue cells (particularly in muscle myoglobin, and nerve). There are many different hemoglobins with different proteins and similar hemes.

Henry's law empirical relation stating that the vapour pressure of a substance is linearly proportional to its concentration in solution. The constant of proportionality is known as the Henry coefficient. The validity of Henry's law is limited to dilute solutions.

herbicide a →pesticide to combat plant pests; a chemical substance for killing plants, especially weeds.

herbivore organism that eats plants. In the sea, herbivores feed predominantly by grazing on →phytoplankton. Most zooplankton and zoobenthos species are basically herbivores, though certain amounts of →detritus and →bacteria ingested with the algae make a clear distinction of true herbivores difficult. See also →grazing.

hermaphrodite animal producing both male and female reproductive cells.

Heterokontophyta see →algae.

Heteropoda superfamily of the order Mesogastropoda, class →Gastropoda (phylum →Mollusca), →pelagic →prosobranch →gastropods with fin-like foot and reduced shell; Examples: *Atlanta, Carinaria*.

heterotrophs organisms taking organic matter from their environment, because they are not able to synthesize organic matter themselves as are their counterpart: the →autotrophs. Heterotrophs are in general all animals and bacteria; autotrophs are plants.

hiatus break or interruption in the continuity of the geological record. The term indicates both (a) the geological time lapse not represented by rocks and (b) the absence of rock units in a stratigraphic sequence of rocks (as a result of either nondeposition or erosion).

Hirudinea class of the phylum Annelida (segmented worms), known as leeches. They occur in freshwater, on land, and in the marine environment, and only part of the ca. 500 species are ectoparasites ("blood-suckers"). The smooth body has 34 anatomical segments (subdivided into annuli) and is flattened, tapered at the front side; there are suckers at both ends. They are →hermaphrodite, eggs are laid in cocoons, in which the embryos develop.

holistic approach the idea that the phenomena in the living world are controlled by the whole of all living organisms. In this, the whole is not merely the addition of the parts, it has properties that cannot be explained from the properties observed in single individuals. Famous is the Gaia hypothesis (1979) of J. Lovelock, which states that all living species together control the physical and chemical conditions in the hydrosphere and the atmosphere in such a way that the circumstances remain suitable for life.

Holocene see →geological time scale.

Holocephalii subclass of Chondrichtyes, see →Pisces.

holoplankton see →plankton.

Holothuroidea sea cucumbers, class of the phylum →Echinodermata. As in the echinoids, the body has no arms. Mouth and anus at opposite sides. The body is five-sided symmetrical with a much prolonged axis. Skeletal plates or ossicles in the skin are microscopic. There are rows of podia on the body, and a circle of branched tentacles round the mouth with which fine food material is collected from the water or the sediment. The sea cucumbers lie on the seabed (bottom dwellers) or live buried in the mud, and a few are →pelagic deep-sea forms.

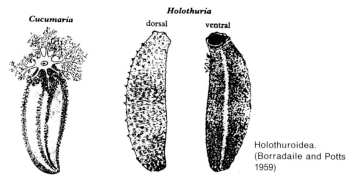

Holothuria

Cucumaria   dorsal   ventral

Holothuroidea.
(Borradaile and Potts
1959)

homeostasis tendency of →ecosystems to correct changes and to remain in a state of equilibrium by negative feed-back responses on density fluctuations in numbers.

homeotherm(ic) (syn. homoiotherm(ic)) warm-blooded animal or more precisely an organism that can maintain a constant body temperature at varying environmental temperatures. In the sea homeothermic animals are scarce: this property is confined to mammalian species and seabirds, which, by means of a thick thermal insulation or feathers, can reduce heat loss and maintain a high, constant body temperature. The internal body temperature of some large fish species may, especially during the heavy exercise of fast swimming, also rise above that of the surrounding water, but, as the body temperature does not remain constant, they are strictly speaking not homeotherms. Compare →poikilotherm.

homing an oriented movement of an animal towards a specific site used for breeding, resting, hibernating, etc. Feeding excursions of invertebrates such as limpets (*Patella*), periwinkles (*Littorina*), and fiddler crabs (*Uca*), often end at the starting point. The mechanism of orientation is not always clear. There are several alternative hypothetical cues such as orientation to gravity, the →earth magnetic field, celestial cues, the sun, or a form of orientation based on smell (olfaction) or memory of past movements (kinesthetic memory), while in intertidal snails retracing of mucous trails has been observed.

homoiotherm(ic) see →homeotherm(ic).

Hoplocarida mantis shrimps, superorder of the class →Malacostraca, subphylum →Crustacea (phylum →Arthropoda). Predators of fish, crabs, shrimps and →mollusks. The second pair of thoracic appendages is enormously developed for raptorial feeding. Mostly tropical, but also in the north Atlantic (e.g., *Squilla*). Live in burrows in the seabed, which they leave when feeding.

horizon, (reflecting) major layer of seismic reflection of the ocean floor. Three horizons are distinguished: horizon A is the uppermost reflecting horizon; horizon B is the lowermost reflecting horizon and horizon β is a horizon occurring between horizons A and B. See also →reflection shooting.

host-parasite relationship see →parasites.

hot spot a localized region on the earth's surface where the →geothermal heat flow is unusually high. Hot spots spring from the →mantle and are often inferred from volcanic activity in the →lithosphere above. When lithosphere drifts over the spot (see →plate tectonics) a succession of locations are uplifted by heating. This process produces a chain of islands (island chain), commonly bearing active or extinct volcanoes on top. A prominent example is the Hawaii-Emperor chain, where volcanism at Hawaii marks the present-day location of the hot spot.

hot spring see →hydrothermal vents.

humic acids marine humic acids and →humic substances are polymerized compounds, containing parts of →amino acids, sugars and →fatty acids with surface-active properties. They are complex, brown, mixed compounds with melanoid structure, and contribute to so-called →Gelbstoff. Part of the marine humic acids may be produced →in situ, but in coastal regions they are predominantly of terrestrial origin.

humic substances major organic constituents of soils and sediments. In seawater they partly determine, for instance, the solubility of substances, sorption, and precipitation of numerous elements, and the optical properties of water. They exhibit no specific physical and chemical characteristics (e.g., a sharp melting point, or an exact elementary composition). Humic substances are dark-colored, acidic, in seawater predominantly aliphatic, hydrophilic, chemically heterogeneous and complex, structurally complicated, poly-electrolyte-like substances. Molecular weight from a few hundred to several thousand. According to their behavior in aqueous solutions at certain pH values they are divided into →fulvic acids, →humic acids and humic substances. The genesis of humic substances and the related →Gelbstoff occurs during the condensation of carbohydrates with →amino acids. In seawater humic substances probably also contain remnants of lipid biopolymers.

hummock pile of fractural sea ice, see →ice.

hydraulic piston corer (HPC) see →corer.

hydrocarbons group of organic compounds containing only two elements, hydrogen and carbon. On the basis of their structure they are divided into aliphatics and aromatics, the former being subdivided into alkanes, alkenes, alkyns, and their cyclic analogs. The simplest hydrocarbon is methane. In seawater hydrocarbons occur in concentrations ranging from 1 to $50 \, \mu g \, dm^{-3}$; in the surface microlayer much higher concentrations have been observed. The content of superficial sediments is in the order of 1 to $100 \, \mu g \, g^{-1}$ of dry sediment.

hydrocast oceanographic sampling by the lowering of a series of sampling bottles into the water column and retrieval of water samples at an →oceanographic station.

Hydrocorallina order of the class →Hydrozoa (phylum →Cnidaria). The ectoderm lays down an →exoskeleton consisting of calcareous grains. The colony has two types of polyps: the gastrozooids with a mouth surrounded with tentacles which nourish the colony, and the dactylozooids with capitate tentacles which catch prey. There are also →medusae with a very short free-living stage, formed by budding. The forms included in this group are mostly associated with reef →corals in tropical seas.

hydrodynamics in hydrodynamics (or fluid dynamics) the effect of mechanical forces on a moving fluid is investigated. In such studies the fluid is considered a continuum: the interaction between the molecules is not taken explicitly into account, but implicitly by the introduction of physical properties like viscosity or surface tension. In principle, hydrodynamics is based upon three equations: the equation of motion, in which the combined effect of all forces acting on elements of the fluid is considered, the continuity equation in which the condition is formulated that the total mass is conserved and the →equation of state, relating mass and volume. The most elementary expression of the equation of motion is that of the Navier-Stokes equations. For practical problems simplified equations are often used.

hydrogen sulfide a very poisonous gas. It is an intermediary of the microbial →sulfur cycle. In aquatic systems it is in a pH-dependent equilibrium with its dissociated forms: $H_2S \Leftrightarrow HS^- + H^+ \Leftrightarrow S^{2-} + 2H^+$.

hydrogen-ion or proton (H^+). These do not exist as such in aqueous solutions, but are associated with water molecules to hydronium ions (H_3O^+). The negative logarithm of its concentration is called →pH.

hydrography stands for two types of description of the seas. In the first place it is used for the charting of the seas, with depths and coastlines etc., primarily for use in navigation, and can be considered marine cartography. However, usually the collection and distribution of other navigational information, including tides and currents, also fall under hydrography in this sense. In the second place it stands for "descriptive oceanography" and is mainly concerned with the description of water properties, their distribution and variations.

hydrological cycle the cycle of water around the globe with the ocean as the largest reservoir with $1.4 \cdot 10^6$ km^3 of water, in which its residence time is 44 000 years. Locally, hydrological cycles indicate the cycle of precipitation, evaporation, inflow, and outflow of water.

hydrolysis chemical reaction in which H_2O is added to the original molecule. Due to hydrolysis organic polymers such as polysaccharides, proteins and lipids are split into their basic units. In the hydrolysis of metal compounds and ions polynuclear hydroxo-complexes are formed that largely determine the behavior of these metals in seawater.

hydromedusa see →Hydrozoa.

hydrostatic approximation approximation of vertical component of the equations of motion in →hydrodynamics, assuming no vertical motion. The pressure distribution in the ocean is derived from the hydrostatic equations derived from this approximation.

hydrothermal vents sites where hot water is ejected. Hydrothermal vents occur both on land and at the sea floor. The existence of such submarine vents had been postulated in the late 1960s. Percolation of seawater through the oceanic →lithosphere upon its emanation along the axes of →mid-ocean ridges was supposed to account for its observed rapid cooling. In the late 1970s the tracing of ambient seawater anomalies (notably for helium

→isotopes, temperature and manganese concentration), combined with deep dives of R.V. submarine *Alvin* (Woods Hole Oceanographic Institution), led to the discovery of several active vent sites in the Pacific Ocean, later followed by vents discovered on the Mid-Atlantic Ridge in the mid 1980s. The emanating vent water has temperatures up to about 350 °C, and a chemical composition very different from normal seawater. Biological communities living around the vent sites are unique, not only because of several new species found, but mostly because they are ultimately driven by geothermal rather than solar energy.

Hydrozoa (hydrozoans) class of the phylum Cnidaria, with small and plant-like colonies of polyps, and with small and fragile jellyfishes as generative stages, called hydromedusae. In most hydrozoans, the polyps form colonies with individuals attached to the central hydrorhiza, which creeps over the substrate. Hydrozoans feed on fine suspended material collected with the tentacles or, as hydromedusae, generally on zooplankton. The Hydrozoan order →Siphonophora contains the remarkable floating colony of differentiated polyps known as the dangerous Portuguese man-of-war, *Physalia*.

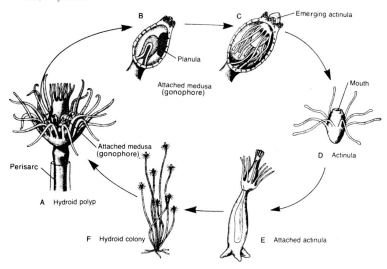

Hydrozoa. (Barnes 1987)

hypersaline waters with salinity considerably higher than normal seawater. Examples are the →brines at the bottom of the Red Sea, and brines in narrow depressions in the sea floor of the East Mediterranean, e.g., the Tyro and Bannock Basins. See also →salinity crisis.

hypertonic adjective to denote that a solution has a higher osmotic concentration than that with which it is compared. It is a useful expression to indicate the direction of passive water transport by osmosis: a hypertonic solution will gain water from the adjacent →hypotonic solution.

hypocenter an →earthquake focus; the point at which an earthquake originates. The projection on the surface of the earth is the →epicenter.

hypolimnion the waterlayer in lakes below the seasonal →thermocline.

hypotonic adjective to denote that a solution has a lower concentration of dissolved solutes than those present in another solution. When two fluid compartments are adjacent to each other the space with the hypotonic concentration will lose water and (if the separating barrier is permeable) gain solutes (see also →hypertonic).

hypsographic curve (hypsometric curve) cumulative frequency profile representing the statistical distribution of the absolute or relative areas of the earth's solid surface at various elevations above, or depth below, a given datum, usually sea level. The curve is bimodal with the →modes at $+100$ m and -4700 m, which reflects the two different parts of the earth's crust: continents and ocean floor. The upper mode is at low altitude above sea level because of subaerial erosion and sedimentation. The intermediate zone, starting at -200 m and ending at ca. 3500 m, corresponds to the transition between continents and ocean floor (generally at the →continental slope).

I

Iapetus Ocean (syn. Proto-Atlantic Ocean) the sea at about the position of the present Atlantic, named after Iapetus, in Greek mythology the father of Atlas. It separated the Baltic (N. Europe) plate from the Laurentian (N. America and Greenland) plate throughout the early →Paleozoic. The Iapetus Ocean was closed when the two plates collided during the late →Silurian.

ice when water freezes to ice the latent heat of freezing set free is 334 joule per gram water. This is practically the same for freshwater and seawater. Because of its salinity, however, the freezing point of seawater is below 0 °C. In seawater with a typical →salinity of 35 the freezing point is about −1.9 °C. Sea ice consists of a framework of ice crystals encapsulating cells of concentrated seawater; this in contrast to the ice of →icebergs, which does not contain seawater. The structure and mechanical properties of sea ice strongly depend on the age and evolution of the ice. For the description of sea ice a special terminology has been developed on the basis of its external appearance held characteristic of its way of formation, such as ice slush (a layer of thin floating ice plates), pancake ice (more or less rounded ice floes), pack ice (cohesive mass of ice floes), sometimes deformed by external pressure into ridges and →hummocks. Large open areas where the ice regularly breaks open because of differential movement are indicated by the Russian name Polynya.

ice age see →glacial epoch.

ice cap mass of ice covering a highland area or a flat landmass such as an Arctic island. Ice caps outrange the underlying topography and are (by definition) smaller than 50000 km². Compare →ice sheets.

ice rafting transport of rock fragments of all sizes on or within floating ice. See also →glaciomarine sediments.

ice sheet →glacier forming a continuous cover over a land surface of (by definition) more than 50000 km². Ice sheets are of considerable thickness and have the tendency to spread radially under their own weight, being not confined to the underlying topography. During the →ice ages large parts of the northern hemisphere were covered by ice sheets: the Fennoscandian ice sheet covered Scandinavia and parts of Northern Europe, whereas Northern America and Greenland were overlain by the Laurentide ice sheet. Compare →ice cap.

ice shelf sheet of thick floating ice attached to a landmass and bounded on the seaward side by steep cliffs. Is is nourished by snow accumulation and by seaward extension of land glaciers. Ice shelves form along the polar coasts (e.g., those of Antarctica and Greenland), some of them extending several hundred kilometers seaward.

iceberg large piece of glacier detached from an →ice sheet (Arctic regions like Greenland) or from an →ice shelf (the Antarctic ice cap). This difference in origin is reflected in their morphology: Greenland icebergs are irregular, Antarctic icebergs are usually tabular. An iceberg is either freely floating in the sea or stranded on a shoal. The submarine part of a floating iceberg is about nine times larger than the visible part. Smaller pieces of icebergs are called floe-bergs (less than 5 m high) and growlers (almost submerged floe-bergs). See also →sea ice, →land ice.

ichnofossil (syn. trace fossil) structure within or on the surface of a sedimentary rock resulting from biological activity, such as (a) burrows and other excavations (the collective name for this type of structuring is →bioturbation), (b) tracks, trails, footprints, resting marks, (c) evidence of feeding activity, and (d) trails of fecal pellets. Many ichnofossils are restricted to limited environments and can therefore be used for the interpretation of the environment of sediment deposition.

ichthyology all science concerning fish (→Pisces), including systematics, anatomy, physiology, ethology, and ecology.

ichthyophage see →piscivore.

IDMS Isotope Dilution Mass Spectrometry. The most accurate and often also most sensitive method for determination of the concentration of a chemical element in seawater or sediment. A known amount of the element with artificially altered ratio of its stable →isotopes is added to the unknown amount (with known natural isotopic signature) in the sample. Upon extraction and purification of the element, the isotopic ratio of the mixture is determined with a mass spectrometer. From this value and the other known values (amount and isotopic ratio of addition, isotopic ratio and volume of sample) the unknown concentration is calculated.

IKMT (Isaacs-Kidd Midwater Trawl) see →nets.

illite a general term for a group of mica-like clay minerals including →glauconite, consisting of a 2:1 layer (see →clay mineral) with interlayered K ions. Besides K, glauconite also contains Na, Ca, Mg, Fe^{2+}, and Fe^{3+}. The general formula for illites is: $K_{1-1.5} Al_4(Si,Al)_8O_{20}(OH)_4$. Illites are developed under alkaline conditions by alteration of micas, alkali feldspars etc., and are widely distributed in sediments and soils.

illuminance the amount of visible light per unit area falling on a surface. Typical measurement units are the lux (lumens per square meter) and the foot candle (lumens per square foot).

immigration process of migration of individuals of a population towards or into a certain area; opposed to →emigration.

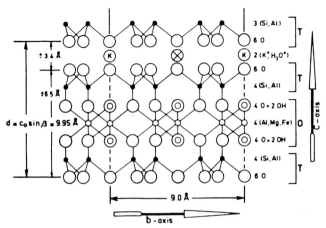

Illite. (After Beutelspacher and van der Marel 1968)

indicator (biol.) a living object (plant, animal, community or parts of it) which gives unequivocal information on (biotic or abiotic) changes in the natural environment, or a change in quality for example, by growth form, behavior, or the accumulation of pollutants. Such changes may result from natural causes or from human activities. Essential for indicators is the reversible relationship between the "dose" parameter and the "effect" parameter. Bioindicators may be local, regional, or usable on a larger scale (e.g., worldwide).

industrial waste see →waste.

inertial oscillation if a fluid on the rotating earth is set into motion (for instance by a sudden impulse), and no horizontal pressure forces are counteracting the current, the only force acting is the →Coriolis force, if the friction may be neglected. As this force works at right angles (except on the equator) to the current, the fluid will continuously turn around and the water particles describe circles in the horizontal plane. These motions are called inertial oscillations or gyroscopic waves. Their frequency, the inertial frequency, is equal to the →Coriolis parameter. (Figure see p. 140)

inertial subrange a range in the spectrum of turbulence connecting the lower frequency, longer-wave part of the spectrum with that part of the spectrum where the molecular viscosity is able to dissipate the kinetic energy of the motion directly into heat. The lower limit of the length scale of motion is the so-called Kolmogoroff length, the part of the spectrum at smaller scales belongs to the viscosity subrange.

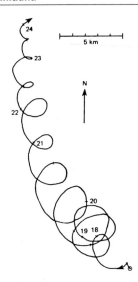

Inertial oscillation. (After Sverdrup et al. 1942)

infauna see →endofauna.

inhibition any process that acts to restrain reactions or behavior. See also →light and →photo-inhibition.

insecticide a →pesticide to combat insect pests; a chemical substance for killing insects.

insects, marine see →neuston.

in situ at the actual spot in the natural environment. Compare →in vitro.

insolation energy from solar radiation reaching the earth's surface. Insolation varies according to latitude, season, and time of the day, as well as according to atmospheric conditions. Part of the insolation over the sea is reflected (see →albedo), part penetrates to greater depths (see →optics). The insolation is a constituent in the →heat balance of the oceans.

instar the stages between molts in →Arthropoda, see also →molting.

intercalibration exercise to compare analytical methods as used in different laboratories on the basis of the analysis of a prepared homogeneous sample, reflecting the matrix (water, dried sediment, dried organic tissues) for which the methods are applied.

interface region between two neighboring phases or compartments in which neither of the two phases can describe its characteristics, e.g., air-sea interface, sediment-water interface.

interglacial time interval between two successive →glacial epochs or stages. These periods are characterized by both the melting of →ice sheets to about their present size, and the maintenance of a warm climate for a sufficient length of time to permit certain large-scale vegetational changes to occur.

internal wave occurs in the interior of the water at the transition between two layers of different density. The frequency is between the →Coriolis parameter and the →Brunt-Väisälä frequency. At such frequencies their speed and wavelength are smaller than those of comparable surface waves, but their amplitudes may be much larger. Internal waves driven by tidal forces are called internal tides. Internal waves manifest themselves by regular variations in the current or in the vertical density structure.

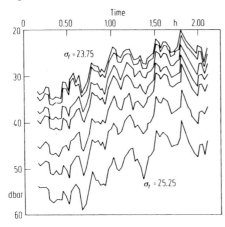

Internal wave.
(Käse and Clarke 1978)

interspecies hydrogen transfer the process in which organic matter is degraded →anaerobically by the interaction of several groups of →microbes in which hydrogen production and consumption are intimately intertwined, e.g., methane-forming bacteria on anaerobic-growing →Protozoa.

interspecific (biol.) between different species, see also →competition and →interaction.

interstitial space the space between sediment grains forms a more or less coherent system of cavities and channels. This interstitial space is filled with interstitial or pore water. The relative size of this space depends on form, compaction, and size distribution of the grains. Sediment grains with a large total surface form a good substratum for all kinds of bacteria and unicellular algae, which provide food for a highly specialized fauna of for

Interstitial space. (Fenchel 1969)

instance →nematode worms, various →Crustacea and →Protozoa. Such interstitial organisms play an important role in the decomposition of organic material, a process usually resulting in enhanced concentrations of nutrients in interstitial water. See also →meiofauna.

intertidal zone see →zonation.

intracellular digestion degradation of food substances into smaller molecules as it takes place inside the cells of living organisms. Usually the enzymatic breakdown of nutrients occurs outside the cells because most digestive enzymes have their →pH optimum at acidic or basic pH values, while cellular pH is always close to neutral. Sometimes, however, food particles are ingested in living cells (→pinocytosis) and digested there. In unicellular organisms the phenomenon is quite common: in higher animals it is only encountered in the epithelial cell lining of the digestive tract.

intraspecific (biol.) within one species, see also →competition and →interaction.

invasions sudden appearance of organisms outside their normal range. The invaders may have used their own means of dispersal for settlement in the new surroundings; they may have been transported by circumstances (e.g., high winds), or unintentionally or intentionally transported by man. In this century European coastal waters have been invaded by several organisms of various origin, such as: the barnacle *Elminius modestus* from New Zealand, the crab *Ericheir sinensis* from China, the bivalves *Petricola pholadiformis* and *Ensis directus* from North America, the alga *Sargassum muticum* and the tunicate *Styela clava* from East Asia. Characteristically, the invaders increase enormously, wiping out entire indigenous communities. Such ecological invasions are only possible when restraining forces are lacking.

Invertebrata the 95 % of the animal kingdom not possessing a backbone (vertebral column). The division of the animal kingdom into vertebrates and invertebrates is artificial and reflects historical human bias in favor of man's own relatives.

in vitro "in glass", in general, applied to biological processes when they are experimentally made to occur in isolation from the whole organism (which usually means within a glass vessel). Compare →in vivo.

in vivo within the living organism. Compare →in vitro.

ion-exchange the process in which cations and/or anions attached to a solid substrate are exchanged for equivalent cations and/or anions from the surrounding solution. The reaction is reversible and can be described with an equilibrium constant, such as →distribution coefficient. In the marine environment, clay-mineral sedimentary particles have some degree of cation-exchange properties, which can be defined by the →base-exchange capacity and determined by the technique of exchange of sodium cations for ammonium cations.

ionic product of water is given by the product of H^+ and OH^- ion concentrations and has a value of 10^{-14}. Thus the pH ($-\log H^+$ concentration) is equal to 7.0 for equal H^+ and OH^- concentrations, where $pH + pOH = 14$. See also →pH.

ionizing the process in which substances dissolved in seawater separate in cations and anions, where each ion is surrounded by water molecules through weak bonds and as hydrogen bonds. The ionic radius of these hydrated ions determines the physical properties of ions in water, like their velocity to pass through membranes.

iridium-rich level thin lamina of sediment (varying from 2 to 15 mm) highly enriched in iridium, which marks the extinction level at the →Cretaceous-Tertiary boundary. The iridium has an extraterrestrial origin; an important aspect of this level is that it respresents a globally synchronous event. See also →extinction and →biostratigraphy.

irradiance the amount of radiant energy per unit area falling on a surface. Irradiance is typically measured in units of watts per square centimeter. In the sea the surface is usually taken horizontally, and we can find a downward and upward irradiance.

irrotational flow see →potential flow.

Isaacs-Kidd Midwater Trawl see →nets.

isentropic analysis see →isentropic surface.

isentropic surface a surface in a fluid where the →entropy is constant. It is also called a neutral surface. In the ocean transport along isentropic surfaces is generally thought to be the main component of lateral turbulent transport of heat and salt. Therefore isentropic analysis, i.e., analysis of the distribution of properties in such a surface, is often used to study the spreading and mixing of →water masses. In first approximation, surfaces of constant potential →density relative to an appropriate pressure are used as isentropic surfaces.

island arc a chain of islands bordered on one side by a deep-sea →trench. Usually it fringes a continent, is convex towards ocean and trench, with a →marginal basin on the concave side. Island arcs are commonly volcanic in origin and often include lines of active volcanoes that then are denoted as volcanic arcs. In →plate tectonic contexts, island arcs occur at the margin of an overriding oceanic plate and are surface manifestations of the subduction process. Island arcs are classified as active or inactive when the subduction process is active or has ended, respectively. Active island arcs are sites of severe earthquake activity. The Aleutians, the Kurils, and the Marianas are typical examples. See also →volcanic arc.

island chain see →hot spot.

iso- prefix, which means equal; in mapping, a curve along which, or a surface upon which, a selected parameter is invariant.

isobar line drawn on a map through places having the same atmospheric pressure at a given time.

isobath line drawn on a map through places having the same depth with respect to the mean →sea level (or a reference level approximating the sea level).

isohaline line drawn on a map through places having the same salinity.

Isopoda (isopods) order of the class →Malacostraca (subphylum →Crustacea, phylum →Arthropoda). Isopoda have no carapace, and not stalked but sessile eyes. The body is usually depressed. Example: the shore slater *Ligia*. (Figure see p. 145)

isopycnic line drawn on a map joining places of equal atmospheric →density.

isostasy theoretical balance of large portions of the earth's →crust as though they were floating on a denser underlying medium. The Pratt hypothesis assumes density variations so that areas of less dense crust rise above areas

Isopoda. (Barnes 1987)

of denser crust. The Airy hypothesis varies the thickness of crustal blocks of constant density so that, similar to floating →icebergs, thicker parts ride higher and reach deeper. The Pratt hypothesis accounts for the elevation of →mid-ocean ridges, whereas the Airy hypothesis explains the deep-seated roots of mountain ranges. Isostasy accounts for the subsidence of areas of deposition, whereas areas of erosion rise. See also →eustasy.

isotherm line on a surface that connects points where the temperature is the same.

isotonic two solutions are said to be isotonic when the concentrations of dissolved substances in both solutions are equal. The composition of the two solutions may be quite different, as isotony only refers to the number of dissolved particles. If in two solutions these numbers are equal, then no net transport of solvent will take place through a barrier which separates the two solutions.

isotopes nuclides of the same element but with different atomic weight. This is caused by a different number of neutrons in the nuclei, which have at the same time an identical number of protons. Isotope concentration ratios like $^{16}O/^{18}O$ and $^{12}C/^{13}C$ are applied as tools to determine various processes in the sea, where either the ratio is shifting along the process, or one of the isotopes or both are decaying due to the emission of radioactive radiation, as in the ratio $^{12}C/^{14}C$. If isotopes undergo radioactive decay they are called →radioisotopes, if not, they are called stable isotopes.

J

jellyfish free-swimming medusoid stage of →Cnidaria, see →medusa.

JOIDES (Joint Oceanographic Institutions for Deep Earth Sampling) a formal agreement signed in 1964 by five of the United States oceanographic institutions to cooperate in deep-sea drilling. It is the intent that JOIDES prepares and proposes drilling programs in trust for the entire scientific community. Under the auspices of JOIDES the →Deep Sea Drilling Project (DSDP) started in 1968. The aim was to drill the oceanic crust in waters down to 6000 m for scientific purposes and the results have proved amazingly successful.

Jurassic see →geological time scale.

juvenile the life stage between larval stage and the adult stage, characterized by the absense of a reproductive capacity.

juvenile water at the beginning of the 20th century the terms juvenile, connate, metamorphic, and meteoric water were applied to define the different kinds of water coming from the interior of the earth in springs and volcanic gases. Juvenile water has never been part of the hydrological cycle on the surface of the earth. When lithification of sediments sets in, →interstitial waters are called connate waters, and when metamorphosis occurs they are called metamorphic waters. Meteoric waters are waters of recent atmospheric origin. No absolute criteria exist to discriminate between these kinds of waters. The total amount of water from volcanic sources alone could fill the oceans in about 100 million years; therefore only a very small part of this volcanic water can be juvenile. It is questionable if juvenile water has ever been found.

K

kaolinite a group of →clay minerals consisting of a 1:1 layer (see →clay minerals) structure with the general formula: $Al_2Si_2O_5(OH)_4$. Kaolinite contains exchangeable cations, and is derived from alteration of alkali feldspars under acidic conditions.

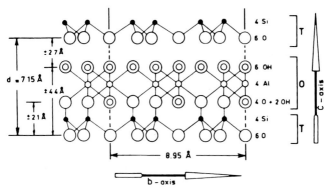

Kaolinite. (After Beutelspacher and van der Marel 1968)

Kastengreifer see →box corer.

katadromous see →catadromous.

kelp see →Phaeophyceae.

Kelvin wave long wave in which for each moment the →Coriolis force is compensated by the instantaneous pressure gradient perpendicular to the direction of propagation, described by the British physicist W. Thomson (Lord Kelvin, 1824–1904) in 1879. It gives a wave with an amplitude that increases perpendicular to the direction of oscillation of the water particles. The increase is to the right in the northern, and to the left in the southern hemisphere. A Kelvin wave should then be bounded by a coastline to the right, or the left, respectively. Along the equator Kelvin waves may also occur, but then bounded by the variation of sign of the →Coriolis parameter. They are traveling eastward. Double Kelvin waves are found in places with an abrupt transition in ocean depth, such as a shelf break. They occur on either side of such a transition, and may have different shapes in the direction perpendicular to the propagation. (Figure see p. 148)

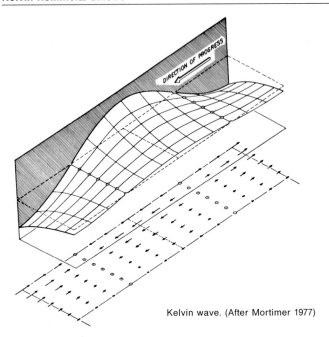

DIRECTION OF PROGRESS

Kelvin wave. (After Mortimer 1977)

Kelvin-Helmholtz billows a wave-like pattern that develops on a →shear layer because of →shear instability, and that is effective in vertical mixing of stratified water.

kinematics part of the mechanics dealing with the movements of bodies; in the ocean, the description of the current pattern by means of current fields, streamlines, current roses, etc.

kinesis general undirected motion of organisms to avoid and escape unfavorable environmental conditions.

Kinorhyncha phylum of small microscopic →Aschelminthes that live in the →interstitial spaces between sediment grains in shallow areas, and feed on diatoms. The chitinous skin or →cuticula is divided into marked segments or zonites. Dorsal cuticular plates bear spines. (Figure see p. 149)

knot measure of a ship's velocity; 1 knot equals 1 →sea mile per hour.

Kolmogoroff length see →inertial subrange.

krill planktonic crustaceans of the order of the →Euphausiacea, that occur in dense swarms in the Antarctic and subantarctic seas and form the main food for baleen whales, pinguins and seals. An important species is *Euphausia*

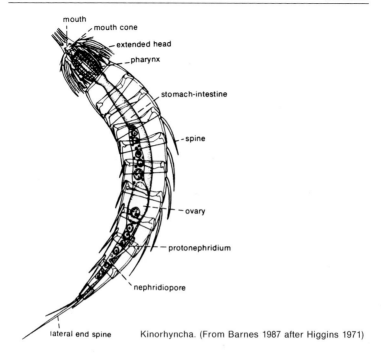

mouth
mouth cone
extended head
pharynx
stomach-intestine
spine
ovary
protonephridium
nephridiopore
lateral end spine

Kinorhyncha. (From Barnes 1987 after Higgins 1971)

superba. Since krill has a high protein content, commercial krill fisheries are developing in Antarctic waters.

K-strategy a →life strategy optimally geared to living in a stable habitat with a high level of interspecific competition. Often, significant energy is invested in defence mechanisms. Parental care is facilitated by low fecundity (small litters of large-sized offspring), by longevity and size. The consequent lower mortality of young leads to more efficient use of energy resources. The birth rate in K-strategists is very sensitive to population density, and will rise rapidly if density falls. K-strategists are unlikely to be well adapted to recover from population densities significantly below their equilibrium level (K) and may become extinct if depressed to such low levels. These organisms, rather than the →r-strategy selected species, need the concern of the conservationist.

Kuroshio strong →boundary current in the western part of the northern Pacific. It flows along the east coast of Taiwan and Japan up to about 35°N latitude, where it turns to the east.

L

labial palps folds in front of the mouth that take part in the collection or selection of food; in, e.g., →Bivalvia.

LAD last appearance datum, see →biohorizon.

lag phase generally the phase between stimulus and effect, more specifically the period after inoculation of a population of microorganisms before growth begins.

lagoon shallow stretch of seawater (channel, sound, bay, salt-water lake) near the sea and/or communicating with it, and partly or completely separated from the sea by a low, narrow, elongate strip of land (reef, →barrier island, sandbank, (sand)→spit).

Lagrangian representation representation of the motion of a fluid by noting the movement of the fluid elements along their trajectories. See →Euler-Lagrange transformation.

Lamellibranchia (lamellibranchs), see →Bivalvia.

laminar flow flow that is not turbulent, but in which the friction is caused by (molecular) viscosity.

land ice any ice formed on land, even though it may later flow into the sea like →icebergs. Compare →sea ice.

Langmuir circulation a type of small-scale circulation observed in the upper layers of seas or lakes, named after the physicist Irving Langmuir, who studied this phenomenon around 1930. It consists of alternate left- and right-hand-rotating helical roll vortices more or less aligned in the direction of the wind. This gives a succession of horizontal zones of confluence and diffluence that become visible because at the surface confluence zone

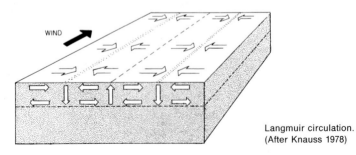

WIND

Langmuir circulation.
(After Knauss 1978)

drifting material (Sargassum weed, oil, foam) is collected. The row spacing ranges from some meters to hundreds of meters. The cause of this circulation pattern is not yet fully clear.

Langmuir isotherm see →Freundlich and Langmuir isotherm.

Larvacea class of the subphylum →Urochordata (phylum →Chordata), also called Appendicularia. Planktonic species consisting of a small body, a long tail and a relatively large balloon of jelly, punctured with small holes, in which they live. This punctured balloon is used as filtering apparatus for →nanoplankton. Historically of interest because study of this filtering apparatus led to the discovery of many previously unknown nanoplanktonic organisms. A typical example of this class is *Oikopleura*.

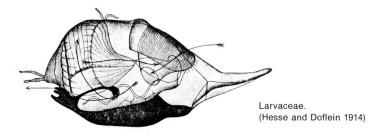

Larvaceae.
(Hesse and Doflein 1914)

larva juvenile form during animal development, differing in shape and appearance from the adult. Larvae undergo metamorphosis before they reach the adult form.

last appearance datum see →biohorizon.

latent heat energy needed for a phase transition from ice to water or from water to water vapor, or energy released in the reverse process. Latent heat is a form of enthalpy (see →thermodynamics). Evaporation from the sea surface often contributes considerably to the →air-sea exchange of heat.

lateral line system system of sense organs, present in aquatic vertebrates (fish and aquatic Amphibia) in pores and canals arranged in a line down each side of the body, and in complicated patterns of lines on the head. Detects pressure changes including vibrations (low-frequency sounds) in water.

Laurasia the late Mesozoic continent from which the present Eurasian and North American continents have been formed. See also →continental displacement, Pangea.

Laurentide ice sheet see →ice sheet.

lava →magma that has streamed out from the crust into the earth's surface, or rock that has solidified from such magma.

Law of the Sea (LOS), a set of rules, established by international treaties and international custom, governing the rights and duties of states (and individuals) concerning the sea. The LOS has been developed and codified during three international conferences, →UNCLOS I, II and III. The last conference (1982) resulted in a new convention, entitled United Nations Convention on the Law of the Sea, abbreviated to →UNCLOS III, which has been signed (up to 1989) by 155 states and ratified by 42 states. Though still short of the required number of ratifications (60), certain rules of UNCLOS III are already binding upon the states through their customary force.

layer 1, layer 2, layer 3, layer 4 the study of →seismic velocities recorded under the ocean reveals layers of increasing velocity. Rock types are characterized by such velocities. The upper layer (layer 1) is the sedimentary layer on top of the sea floor. Below the sediments, the →basalts (layer 2) and →gabbros (layer 3) compose the →oceanic basement. Below, the →Moho layer (4) is encountered; extra-heavy rocks called ultramafics are required to produce the very high velocities observed in this layer.

layer model in numerical →models of ocean dynamics in principle the equations have to be solved for three directions in a three-dimensional grid. However, for many problems the vertically integrated equations can be used in a two-dimensional (horizontal) model. To bridge the gap between the two- and three-dimensional approach, the ocean is sometimes divided into a number of coupled layers, for each of which a two-dimensional approach is applied.

leaching gradual release of adsorbed solutes from sediment or soil grains to water that percolates through the sediment or soil.

lee wave originally used in meteorology, where this term refers to waves observed in the lee of mountains; a lee wave in oceanography is a nonhydrostatic internal or surface wave generated by a flow at the lee side of an obstacle, or behind a disturbance that moves with a speed exceeding the →phase speed.

Leptostraca order of the class →Malacostraca, subphylum →Crustacea (phylum →Arthropoda). Small marine crustaceans, the most primitive of the Malacostraca, with eight instead of six abdominal segments. They are shallow-water suspension feeders. Example: *Nebalia,* a ca. 12-mm-long crustacean living on muddy bottoms.

level of nomotion to calculate the absolute velocity of currents by means of the →geostrophic balance a reference level at a certain depth is needed where the →isobaric plane can be assumed to be horizontal. In this plane there should be no current: the level of nomotion. As there are no indirect ways to establish this level for certain (if it exists at all), the choice of it is a matter of judgement.

Liebig's law of the minimum law, formulated by A. Mayer in 1900, based on observation by J. von Liebig of 1840, who stated that harvest in agriculture was limited by that nutrient which was in the minimum, and that by addition of that particular nutrient the harvest could be enlarged. See also →limiting factors.

life cycles of animals the various phases through which an individual species passes from birth to maturity and death.

life strategies the ecological strategies that are evolved to maximize the survival and fitness of the organism in its environment. These strategies for any organism lie somewhere along an *r-K* continuum (*K* refers to the →carrying capacity and *r* to the maximal intrinsic rate of natural increase r_{max} in the →logistic equation). The characteristics of an organism's habitat in relation to its size and range will largely define its optimal strategy. A very important habitat characteristic is its duration stability: the length of time a particular habitat type remains in a particular geographical location. This may range from hundreds of years on a coral reef to a week in a phytoplankton bloom. Clearly the significance of this duration stability depends on the relationship between the organism's generation time (t) and the length of time the habitat remains favorable (*H*). In species where the ratio t/*H* approaches unity, one generation cannot affect the resources of the next; there will be no evolutionary penalty for overshooting the carrying capacity of the habitat. These species then are selected for an →r-strategy. Conversely, for animals that occupy long-lived habitats where the carrying capacity (*K*) is fairly constant, significant overshooting will lower *K*, and will adversely affect subsequent generations. Since many other species will have colonized such stable habitats, interspecific →competition is likely to be intense. Such species are selected for harvesting food efficiently in a crowded environment and their optimal life strategy is a →K-strategy.

life tables the numbers or densities (numbers per unit of area) of different developmental stages are tabulated in life tables, representing the population state in subsequent generations. Numerical techniques have been developed to derive from such life tables insight into key factors, stability, regulation, and the causes of density fluctuations. See also →population dynamics.

light the sun is the only noteworthy source of light for the earth but the amount of solar radiation varies from latitude to latitude. The amount of solar radiation is usually measured with a solarimeter, having a wavelength range from 300 to 3000 nm. Marine phytoplankton uses only radiation in the range from 400 to 700 nm. The action spectra, i.e., relative absorption of various wavelengths by intact →algae differ and strongly depend on the presence of various →chlorophylls and accessory pigments. The depth to which the radiation used for →photosynthesis penetrates (→photic zone) is limited and varies with the wavelength (see →optics). Light in the ultraviolet region (200 to 400 nm) penetrates only in the upper few meters. The depth to which only 1 % of the surface illumination penetrates is called the euphotic zone. Also the attenuation of the various wavelengths in the photosynthetically active region varies. The red wavelengths penetrate least and the blue/green deepest. This situation is reversed when there are dense phytoplankton populations with a high chlorophyll concentration absorbing in the blue region. →Phytoplankton photosynthesis is a function of light intensity, it increases with light intensity to a maximal value when algae become light (I) saturated (P_{max} or assimilation number). At higher irradiation levels the P versus I curve may show a decrease again due to the so-called light inhibition or photo-inhibition effect.

light compass reaction movement of organisms at a constant angle to parallel rays of a light source. In nature this light source is usually the sun and it is called sun compass reaction or orientation.

light inhibition see →photo-inhibition.

limiting factors resource element of food or nutrients which is at a concentration too low to allow maximum growth or reproduction of a dependent organism. In the simplest concept according to →Liebig's law of the minimum growth is limited by the one resource element, which is in the minimum.

limits of tolerance concept in physiological ecology concerned with the tolerance of organisms to environmental factors. An example is the range of temperatures over which a species can reproduce or survive and which shows a lower and upper limit of tolerance. Since organisms often perform differently in different phases of their life cycles, it is sometimes problematic to measure their performance, and the upper and lower values of tolerance of an organism to the range of environmental factors.

limnetic zone that part of the open water in freshwater lakes which is above the oxygen compensation level and thus within the →euphotic zone. The deeper part of the open water of lakes is called profundal zone.

Linnaeus, Carolus (Carl Linné) see →taxonomy.

lipids naturally occurring substances characterized by their insolubility in water and their solubility in organic solvents (fat solvents) such as ether, chloroform, boiling alcohol and benzene. Chemically, the lipids are either esters of →fatty acids or substances capable of forming such esters. They are widespread in nature, being found in all vegetable and animal matter. Some members of this group, such as the phosphatides and sterols, are found in all living cells where, with →proteins and →carbohydrates, they form an essential part of the colloidal complex of cytoplasm. Complex lipids are also found in large quantities in brain and nervous tissues. Other lipids, such as fats and oils, represent the chief form in which excess nutrients are stored in the animal body. They arise from ingested lipids and from the metabolism of carbohydrates and proteins, and are stored in fat deposits, such as subcutaneous connective tissue. Lipids act as heat insulators and as reserve supplies of energy.

lithofacies see →facies.

lithosphere the outermost rigid layer of the earth, commonly 100 km thick, including the →crust and the uppermost part of the mantle. It is broken into →plates and rests on the more yielding →asthenosphere. See also →earth's internal structure.

lithostratigraphy (syn. rock stratigraphy) element of stratigraphy that deals with the lithology of strata and with their organisation into units based on lithologic character. This lithologic character is also an important tool for →facies interpretations: the composition, structure and texture of sedimentary rocks can be used to determine the depositional environment (see →sedimentary structures). See also →stratigraphic hierarchy.

lithotroph see →metabolic diversity.

littoral zone zone between tide marks, i.e., between high-tide and low-tide level. See also →zonation and →coast.

logarithmic profile refers to the velocity profile, and is a function representing the logarithmic variation of flow velocity with the distance from a rigid boundary. It gives an adequate description of the profile for the part of the →boundary layer where the →stress can be considered to be constant. The velocity at a given level is proportional to the so-called friction velocity, which is the square root of the surface stress divided by the density and the nondimensional Von Kármán constant k (about 0.41). The distance from the boundary is scaled with the roughness length of the underlying surface, a measure of the actual irregularity of this surface.

logging see →well logging.

logistic equation the mathematical expression for the logistic equation is $dN/dt = rN(1 - N/K)$ in which N describes a →population of one species, r measures the intrinsic per capita growth rate, and K the total →carrying capacity of the community of which N is a part. This is the simplest differential equation with two very important features which often occur in describing the dynamics of a population: (a) when N is small the equation is reduced to an exponential equilibrium; (b) as t increases, N approaches a steady state value K without oscillations.

long wave a →gravity wave with a wavelength that is large compared with the depth. This may apply to sea waves approaching the beach or to →tides in the deep ocean. They may also be called **shallow water waves**, although this may appear inappropriate in the latter case. Waves with a wavelength

20 times the water depth are considered long waves. Their phase velocity (c) equals the group velocity and is given by $c = \sqrt{(gD)}$, where g is the acceleration of gravity and D is the depth. Differences in depth cause variations in the velocity and long waves are therefore →refracted by topography. For many long waves, having wave lengths of the →deformation radius, the effect of the →Coriolis force cannot be neglected, and this causes different types of waves such as →Kelvin waves and →Sverdrup waves.

longevity the natural lifetime of an organism, which shows a strong positive correlation with its size at maturity. This is probably due to longevity being inversely proportional to total metabolic activity per unit body weight and to the fact that the smaller the organism the higher the level of this activity.

Longhurst-Hardy recorder see →continuous plankton recorder.

longshore current see →coast and also →wave-induced currents.

longshore drift see →coast.

lophophore a food-catching tentacular organ present in four phyla (→Bryozoa, Entoprocta, Phoronida, and →Brachiopoda), often grouped together as the Lophophorates. It is a circular or horseshoe-shaped fold of the bodywall, that encircles the mouth and bears numerous ciliated tentacles.

Lotka-Volterra equations set of differential equations which constitute a model for a deterministic one-predator-one-prey system with continuous growth. The set of equations is: $dH(t)/dt = H(t)[a - \alpha P(t)]$; $dP(t)/dt = P(t)[-b + \beta H(t)]$. $H(t)$ and $P(t)$ are the populations of prey and predator, respectively, at time t. The parameter a relates to the birth rate of the prey, b to the death rate of the predator and α and β to the interaction between the species. This set of equations constitutes the simplest representation of the essentials of a nonlinear predator-prey interaction.

luciferase see →bioluminescence.

luciferine see →bioluminescence.

luminance the amount of visible light per unit area per unit solid angle emitted by a light source. Typical measurement units are candela per square meter and foot lambert.

luminescence see →bioluminescence.

lunar cycle a biological →rhythm with minima or maxima once or twice every lunar month (29.5 days), at the time of a certain moon phase. Examples are mating of several coral reef fishes, and release of gametes or larvae by invertebrates such as →Polychaeta, sponges or corals. Since tidal cycles run parallel with lunar phases, it is difficult to distinguish among relevant triggers such as moonlight or hydrographic phenomena.

lunar day mean period between two successive passages of the moon through a certain meridian. As for the →tides the effect of the moon predominates over that of the sun, the lunar day (24.84 h) determines the periodicity of the diurnal and semidiurnal tides.

lunar tide tide caused by the interaction between the moon and the rotating earth (see →tides). The lunar tide predominates over the solar tide notwithstanding the larger mass of the sun, because of the proximity of the moon to the earth. The most important lunar tidal constituent is indicated by M_2 (see →partial tide) and has a period of 12.43 h.

lysocline see →carbonate dissolution.

lysosomes cell organelles containing digestive enzymes, the lysozymes.

M

macrobenthos bottom-living organisms larger than 1 mm, see →benthos.

macronutrients those nutrients that constitute the main part of nutrition, such as organic molecules, →phosphate, nitrate and →ammonium.

macrophytes the larger, macroscopic, multicellular plants as opposed to single-celled, microscopic microphytes.

macroplankton large-sized (between 2 and 20 cm) planktonic organisms, see →plankton.

Madreporaria (syn. Scleractinia) order of the class →Anthozoa (phylum →Cnidaria). Stony →corals. Solitary or colonial, with a heavy, external calcareous skeleton. They have relatively small →polyps. Mainly forming colonies in tropical seas (coral reefs), but also solitary and in cooler regions.

madreporite see →Echinodermata.

magma a hot silicate melt within the earth containing suspended crystals and dissolved gases. Rocks originating from the crystallization and solidification of magma are said to be magmatic. See also →lava.

magnetic anomaly any departure from the theoretical earth's →magnetic field. The earth's magnetic field can be locally affected by induced or remanent magnetism of geological features, which then reveal themselves by magnetic anomalies.

magnetic (earth) field the earth's magnetic field is thought to derive from flow patterns in the liquid outer core (see →earth's internal structure). Its mean field intensity at the surface is ca. 0.5 Gauss (equivalent to 50×10^{-6} Tesla). At any point on the earth, the total field intensity can be resolved in the intensity of the horizontal component and the intensity of the vertical component. The direction is described by the deviation from the north (declination) and the angle with the horizontal (inclination). The earth's magnetic field resembles for 90% a dipole field with its axis declined 11° from the axis of rotation. The small additional nondipole field is responsible for irregularities of the field in distribution and time (secular variation). See also →magnetic anomaly and →magnetic polarity.

magnetic polarity the converging of the magnetic lines of force in a dipole, with a magnetic north and a magnetic south pole. According to the convention that the northwards pointing end of a compass-needle is a magnetic north pole, the earth's dipole has its (attracting) magnetic south pole near the geographical north pole and its magnetic north pole near the geographical south. However, this polarity is not permanent. Studies of the remanent magnetism induced by the earth's →magnetic field in ferromagnetic rocks at the time of their formation reveal that magnetic polarity reversals have occurred many times in the geologic past. Based upon these reversals, a polarity time scale has been established.

magnetic polarity reversal see →magnetic polarity.

magnetic tags see →tagging.

magnetometer an instrument used for measuring earth magnetic field intensity. Magnetometers may measure the vertical component of the magnetic field, or the horizontal component sometimes the total field is measured. See also →magnetic (earth) field.

magnetostratigraphy stratigraphy based on paleomagnetic signatures. During sediment deposition or cooling of molten rock, ferromagnetic minerals orientate themselves according to the existing magnetic field. The last 4.5 million years include four magnetic epochs (chrons) from old to young: the Gilbert reversed, the Gauss normal, the Matuyama reversed, and the (present-day) Brunhes normal chron. See also →stratigraphic hierarchy, →magnetic polarity and →magnetic (earth) field. (Figure see p. 160)

magnetotaxis →orientation of organisms possessing magnetic crystals (some bacteria, dolphins, some birds) in the earth's →magnetic field.

Malacostraca class of the subphylum →Crustacea (phylum →Arthropoda), containing most of the Crustacea including the larger crabs, shrimps, and lobsters. For general description of body structure see →Crustacea. The trunk of the Malacostraca is composed of 14 segments of which eight segments form the →thorax (sometimes covered by the →carapace), the remaining six form the abdomen and a →telson. The segments each bear one pair of appendages. On the head are two pairs of antennae and a pair of stalked eyes. In the more primitive forms, all thoracic appendages are similar and used as walking legs; in others the first one, two or three pairs are changed into →maxillipeds. The abdominal appendages are smaller, and adapted for swimming, burying, effectuating water currents for ventilation, or carrying eggs (in females). The Malacostraca include the orders →Euphausiacea, →Decapoda, →Mysidacea, →Cumacea, →Tanaidacea, →Isopoda and →Amphipoda.

mammals, marine different groups of originally terrestrial mammals that have returned to the sea over the last 80 million years, probably attracted by abundant food, or to escape from predators or climatic changes. They include cetaceans (whales and dolphins), pinnipeds (seals, sea-lions and walruses), sirenians (manatees, dugongs and Steller sea-cows) and some mustelids (a few otters, and the extinct sea-mink) and sometimes the polar bear is included. Respiratory, circulatory, locomotory, thermo-regulating, and communicative systems are all significantly different from the typical

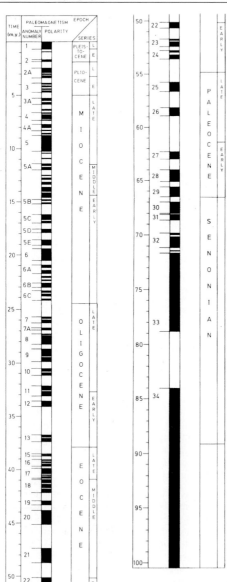

Magnetostratigraphy.
(After Lawrie
and Alvarez 1981)

terrestrial mammalian pattern. Man has used marine mammals as a source of food, fuel, and clothing for at least 5000 years; with extending size and range of hunting vessels man began to affect populations, some species being hunted to extinction: the Steller sea-cow was discovered in the Bering Sea in 1741 and eradicated about 30 years later, the bowhead and right whales were almost eradicated in the 18th and 19th centuries, the blue whale and humpback in this century.

manganese nodules (syn. ferromanganese nodules spherical or semi-spherical or irregular concretions, largely consisting of oxides of iron (Fe) and manganese (Mn), but containing in addition metals like titanium, chromium, copper, nickel, cobalt, and zinc, often found on the surface of deep-sea marine sediments. Most likely of →diagenetic origin. First reported during the →Challenger Deep Sea Expedition. These nodules form very slowly, over millions of years. See also →mineral resources.

mangroves see →salt marsh.

mantle (1) (biol.) see →Mollusca; (2) (geol.) see →earth's internal structure.

marginal basin (syn. back-arc basin) basin of intermediate (ca. 2000 m) to normal oceanic (ca. 4500 m) depth, behind →island arc systems and separating the →trenches and island arcs from continents or other (usually inactive) →island arcs.

marginal seas more or less enclosed seas along the continents.

Mariana Trench the deepest point in the ocean. At this depth an accurate measurement meets difficulties. The best available estimate is about 11 030 m depth. See →trench, sea-floor topography.

mariculture see →aquaculture.

marine provinces see →biogeography.

marine snow the flocs of suspended particles in the ocean, see also →flocculation.

marine terrace see →terrace, marine.

marsh see →salt marsh.

mass extinction see →mass mortality.

mass mortality catastrophic dying of one or several species of organisms (usually not all species are affected) in a limited area. Mass mortality in the sea often coincides chiefly with →red tides and is then due to →blooms of poisonous algae (often →Dinoflagellata). However, not all red tides coincide with mass mortalities and other causes of mass mortalities in sea are known, such as volcanism, earth- and seaquakes, sudden changes in salinity or temperature. Mass extinctions are large mass mortalities that occurred in the past, and worked over longer periods during which numerous species are exterminated, e.g., at the →Cretaceous/Tertiary boundary, which marks the extinction of, e.g., dinosaurs and ammonites. On the causes of such mass extinctions we can only speculate, but usually worldwide extra-terrestrial causes are suggested. The current extinction of many organisms due to activities of man is probably the largest mass mortality ever. See also →extinction and →biohorizon.

mass spectrometry an analytical technique for the identification and quantification of unknown compounds. It is based on the determination of the mass/charge ratios of ionized molecules and ionized fragments of these molecules. The molecules are fragmentized and ionized by exposure to an excess of energy. The ionized molecules and fragments are then accelerated and fed into an ion-analysis device. Here the ions are separated according to their mass/charge ratios and detected by a spectrum recording system. The recorded fragmentation pattern contains specific and characteristic information on the molecule structure. Mass spectrometers are often combined with other analytical instruments (see →gas chromatography). The combination of gas chromatograph, mass spectrometer and data system is one of the most powerful analytical devices for the analysis of complex mixtures of unknown compounds.

Mastigophora (syn. Flagellata, flagellates) subphylum of the phylum Sarcomastigophora, a protozoan phylum belonging to the kingdom Protista. Unicellular organisms distinguished by the presence of one or more →flagella. There are different →autotrophic groups, together called the Phytomastigophora or phytoflagellates, and →heterotrophic groups, together the Zoomastigophora or zooflagellates. See also →Protozoa.

Matuyama epoch, - chron see →magnetostratigraphy.

maxilliped see →Crustacea.

maximum allowable concentration (MAC) [syn. maximum permissible concentration (MPC)] the highest concentration of any (harmful) substance allowed by regulations to occur and not to be exceeded in any or a particular compartment of the environment. See also →bio-assay for the determination of MAC.

maximum sustainable yield maximum harvest which is currently compensated by the fish production of the system. See also →fishery science.

mean the average (arithmetic mean) value of a set of observations. See also →distribution.

mean sea level see →sea level.

meander originally a term indicating a bend in a river, but is by analogy also applied to large bends in ocean currents, such as the →Gulf Stream.

median the central value in a set of observations. See also →distribution.

mediterranean seas seas that are separated from the ocean by land masses or island chains and only connected with the ocean by more or less narrow straits. Apart from the Mediterranean proper, the Caribbean, the Red Sea, and the Arctic seas are considered mediterranean seas.

medusa (jellyfish) free-swimming reproductive life stage of →Coelenterata. Bell- or umbrella-shaped body of which the rhythmic muscular contractions serve for propagation.

megabenthos large bottom-living organisms, see →benthos.

megalopa →pelagic developmental stage of crabs (see →Malacostraca) which comes after the last larval stage or →zoea. The megalopa has already the appearance of the crab, although the abdomen is not yet folded under the body; it is also called the postlarval stage. After the megalopa stage the postlarvae settle on the seabed and become small crabs.

megaplankton planktonic organisms >20 cm, see →plankton.

megaripple →cross-bedded sedimentary structure (with ripple shape) resulting from current powers higher than those necessary for small-scale ripples. Two types can be distinguished: (a) →dunes, with a sinuous to lunate-shaped crest, and (b) sand waves, with a straight crest. The difference between the two types originates from the method of transport: for dunes, the major mode is transport by traction, for sand waves it is suspension transport.

meiobenthos (syn. meiofauna) group of animals of a size between roughly 0.1 and 1 mm living in the spaces between sediment grains. To this group belong for instance various genera of →copepods, nematodes, polychaetes and →ciliates. See →benthos and →interstitial space.

meiofauna see →meiobenthos.

membrane sheet-like structure separating the cell interior from the outside or separating intracellular spaces from each other. Membranes play a major role in controlling the transport of substances between various cellular spaces. In principle, membranes consist of lipid bilayers. As integral part of the membrane structure, or loosely attached to it, protein molecules may occur which function as carrier molecules for the transport of certain solutes across the membrane.

meristic characters characters in organisms that show repetition as a consequence of segmentation of the animals.

meroplankton temporary plankton consisting of →pelagic stages of organisms which also have →benthic stages. Mainly larvae of sedentary organisms. See also →plankton.

mesenterium (mesentery) (1) vertical partitions in →coelenteron of →Anthozoa; (2) in vertebrates: double layer of peritoneum attaching stomach, intestines, spleen etc. to dorsal wall of peritoneal cavity.

mesh size size of the openings in fishing →nets or plankton gauze, measured according to standard methods. The gauze of plankton nets consists of single nylon fibers of a certain diameter, woven and melted together to form square pores that are stable in form. Mesh size of different plankton nets is given in μm length of the side of the square opening, irrespective of the diameter of the surrounding fiber, and ranges from 20 to 2000 μm with 50, 100, 200, 300, 500 and 1000 μm being the sizes most frequently used in plankton research. The nets used in fisheries are made of a great variety of natural and artificial fibers, which can be woven or knotted in many different ways into the netting. Since the mesh openings are not stable in form, and the fibers are sometimes elastic, mesh sizes are measured in a way standardized by the ICES (see →fisheries research). A measuring device is brought diagonally into the mesh opening and stretches it completely with a certain standard force, to measure the "stretched mesh size", the measure used in the minimum mesh size regulations for fisheries for the protection of young fish.

mesocosms in- or outdoor established experimental ecosystems on a meso-scale, i.e., having a volume of between 1 and 10^4 m^3 in →pelagic systems and a surface area of >1 m^2 in →benthic systems. Microcosms or microecosystems refer to smaller experimental systems. Mesocosms provide realistic facilities for basic marine biological research, ecotoxicological testing, and →aquaculture experiments and bridge the gap between laboratory experiments and field research.

mesogloea the jelly-like more or less structureless mass, filling the space between the →ectoderm and →endoderm of the →Coelenterata.

mesopelagic see →zonation.

mesophile a term referring to organisms living in the temperature range around that of →warm-blooded animals.

mesoplankton planktonic organisms of intermediate body size between 0.2 and 20 mm, see →plankton.

mesosphere see →earth's internal structure.

mesotrophic see →eutrophic.

Mesozoic see →geological time scale.

Messinian see →salinity crisis.

metabiosis cooperation of different organisms to degrade substances, e.g., refractory substances such as lignin or cellulose. These are first attacked by specialist microorganisms possessing enzyme systems for the decomposition of such substances and after that other microbes that can utilize the intermediate products follow. Thus the first create the pre-conditions for the development of the latter group of organisms, by whose activities an accumulation of harmful metabolic products is avoided. This kind of cooperation is widespread in nature.

metabolic diversity the overwhelming amount of morphologically different types of living organisms shows a remarkable unity in their biochemistry. Also in the metabolic processes of evolutionary different groups of organisms an impressive diversity exists. Especially among microorganisms a large variety of metabolic types can be found. On the basis of the energy source or the carbon sources used, organisms can be divided into various catagories. If the energy source is chemical the organisms are chemotrophs. If they use light as energy source they are phototrophs. The chemotrophs can be divided into organisms that use the energy from reduced inorganic compounds (hydrogen, reduced S-compounds, reduced N-compounds, reduced Fe, or reduced Mn) the chemolithotrophs or lithotrophs and organisms that use the energy from organic matter the chemoorganotrophs. The generation of energy must be separated clearly from the carbon source they use for building their biomass. If organisms use carbon dioxide as C-source they are autotrophs and if they use organic matter as C-source they are heterotrophs. So chemoorganotrophs are always heterotroph. But organisms such as the sulfate-reducing bacteria that use hydrogen as energy source and organic matter as C-source belong to the category chemolitho-heterotrophs. The hydrogen bacteria using hydrogen as energy source and carbon dioxide as C-source belong to the chemolithoautotrophs. Phototrophs are mostly autotroph (photolithotrophs or photoautotrophs) but if they use organic matter as C-source they belong to the photoorganotrophs or photoheterotrophs. Metabolic diversity exists also in the way of oxidizing reduced inorganic or organic matter. The electron acceptor can be oxygen, nitrate, iron, manganese, sulfate or carbon dioxide. In this way various →aerobic and →anaerobic respiration principles are possible. There are also different types of →fermentations in the group of heterotrophic anaerobic organisms.

metabolism all the biochemical reactions in a cell or organism. Such reactions involve three major phases: first, the breaking down of organic compounds into smaller fragments (catabolism or destructive metabolism); second, the reactions of the intermediary metabolism; third, the reactions of the synthesis of building blocks and polymers for the composition of the various cell structures. These synthesizing reactions (constructive processes) are called anabolism. The metabolism of all the enzymatic reactions in a wide variety of living organisms have much in common. There is a fundamental similarity in carbohydrate, lipid and aminoacid metabolism, which gave rise of the dogma of the unity in biochemistry. On the other hand there is a wide diversity of reactions (see →metabolic diversity), but they are not so diverse as they may first appear.

metameric repetition of a pattern of elements belonging to each of the main organ systems of the body, along the length axis of the body, or a comparative repetition along the axis of an appendage. Example: →Poly-chaeta.

metamorphic water see →juvenile water.

Metazoa multicellular animals in which there is at least some degree of cell differentiation. Include all taxa of the animal kingdom except →Protozoa.

meteoric water see →juvenile water.

methane bacteria (syn. methanogenic bacteria) a unique group among →prokaryote organisms because they produce methane as a major product of →anaerobic metabolism.

mica a group of platy, white-, green- or brown-colored minerals consisting of a 2:1 layer (see →clay mineral) with the general formula: $(K,Na,Ca)(Mg,Fe,Li,Al)_{2-3}(Al,Si)_4O_{10}(OH,F)_2$. Micas are rock-forming constituents of igneous and metamorphic rocks.

Michaelis-Menten kinetics describe the rate of a biochemical process as a hyperbolic function of the concentration of the converted component. Originally derived from a mechanistic analysis of enzyme-catalyzed reactions by L. Michaelis and M.M.L. Menten in 1913:
$V = V_{max} \, S \, /(K_m + S)$, where V is the actual rate at which the reaction proceeds, V_{max} is the maximum rate, S is the concentration of the converted component (often referred to as: substrate concentration) and K_m is a half saturation constant numerically equal to S at 0.5 V_{max}. Not only applied for enzyme kinetics but also for nutrient-uptake kinetics. At low concentrations the rate of the reaction is determined by the initial slope of the Michaelis-Menten curve. This slope, as S approaches zero, becomes equal to V_{max}/K_m. This ratio can be used to calculate the affinity of an enzym for its substrate. In the application to the kinetics of microbial growth the equation is generally referred to as the Monod equation. With regard to microbial growth, the equation is written: $\mu = \mu_{max} \, S/(S + K_s)$ where the specific growth rate (μ) is a function of the concentration of the growth rate-limiting substrate S, the maximum specific growth rate μ_{max} which is observed when S becomes saturating, and K_s (half saturation constant for growth). The ratio μ_{max}/K_s is a measure of the competitive ability of a →microorganism under conditions at which it has to compete for a limiting substrate at concentrations much lower than K_s.

microaerophilic requiring oxygen at very low concentrations.

microbe see →microorganism.

microbenthos →benthos consisting of the smallest animals, i.e., →Protozoa. The arbitrary upper boundary of the animal size is generally set at about 100 μm, which is the lower limit for →meiobenthos. See also →benthos.

microbial decomposition the return of constituents of organic matter to ecological cycles by the activities of microorganisms, see also →biogeochemical cycles. This recycling of material is essential for life on earth.

microbial loop term describing the complex of organisms and processes thought to be responsible for transferring back into the traditional grazing food chain dissolved organic matter "lost" from phytoplankton either by →exudation, →autolysis, or due to →herbivore feeding.

microbiology study of →microorganisms.

microcosm microecosystem see →mesocosm.

microfauna microscopic animals roughly less than 0.1 mm in length. See also →microbenthos and →microorganisms.

microflagellates a diversity of criteria has been used in classifying organisms as belonging to the microflagellates; photosynthetic cells have been alternately included and excluded. The concept has no systematic value and has been loosely applied to distinguish the smallest →eukaryotic cells, often flagellated, from larger cells. See also →Mastigophora.

microfossil fossil too small to be studied without the aid of a microscope. May be either the remainder of a microscopic organism or small part of a larger organism (examples of the last: →otoliths and →phytoliths). By size we distinguish micro- and nanofossils, the last being only visible at the highest light microscope magnification or with the aid of an electron microscope (e.g., →coccoliths and →discoasters). Several useful groups of microscopic organisms are studied, which can be divided on the basis of skeletal composition: calcareous (foraminifers, coccoliths, discoasters, ostracods, pteropods, calpionellids), siliceous (radiolarians, diatoms, silicoflagellates) and organic-walled (spores, pollen, dinoflagellates, chitinozoa). See →micropaleontology and →biostratigraphy.

micronutrient substance which an organism must obtain from its environment to maintain health, though necessary only in minute amounts, either vitamins or trace elements. See also →trace element.

microorganisms (syn. microbes) organisms of very diverse groups including →viruses, prions, viroids, bacteria, fungi, algae and →Protozoa. The first three groups are all acellular and incapable of independent existence. The study of microorganisms is called microbiology.

micropaleontology the study of →microfossils. The rapid evolutionary changes of microscopic organisms make micropaleontology an important biostratigraphical tool (see →biostratigraphy), while the specific tolerances of microorganisms can be used for paleo-environmental interpretations. Once the relations have been established between the distributions of species and environmental parameters such as temperature, salinity, and water depth, this relationship can be used to reconstruct the geologic history, e.g., the history of the oceans by studying distributions of marine microfossils in the sediment (see →facies). A major advantage of micropaleontology is that only small quantities of sediment are needed: often thousands of microfossils can be found in 1 g of sediment.

microphytobenthos small (usually unicellular) plants living on the (shallow) sea floor. Particularly abundant on →tidal flats. See also →benthos.

microplankton small-sized (between 20 and 200 μm) planktonic organisms, see →plankton.

microtektite small glassy object, usually droplet-shaped and less than 1 mm in diameter, found in some deep-sea sediments, and obviously strongly related to →tektites.

mid-ocean ridge a broad, bilaterally symmetrical, elongate submarine swell with sloping sides, 1 to 3 km in elevation, about 1500 km in width, and over 60 000 km in length. Mid-ocean ridges are formed at divergent →plate boundaries and are not necessarily located in the middle of oceans. The new oceanic floor accreted there by volcanic activity sinks as each side moves laterally away and cools, which gives the ridge its form. Examples: Mid-Atlantic Ridge, East Pacific Rise, and Carlsberg Ridge. See also →plate tectonics.

migration periodical movement of organisms between alternative habitats, e.g., the area of reproduction and one or more areas of nonreproductive life, or between areas of foraging and areas used for other activities. Migrations occur at predictable times; in most cases they anticipate unfavorable conditions and are triggered by stimuli timed by →biological clocks. Organisms capable of migration have a sense of direction (→orientation) and are able to move actively. Movements that do not include an obligatory return journey have been classed as dispersal, e.g., young that leave the area of birth and settle elsewhere (natal dispersal), or animals that have reproduced once in one area and use another area the next time (breeding dispersal). See also →orientation, →homing, and →scattering layer.

Milankovitch theory a theory of glaciation, formulated in 1941 by the Yugoslav mathematician Milutin Milankovitch (1879–1958). This theory states that climatic changes result from fluctuations in the geographic and temporary distribution of insolation, determined by →orbital variations.

mile see →nautical mile.

mimicry resemblance to organisms of other species, either to advertise an effective protective device that they possess in common with other species (Müllerian mimicry), or to "pretend" they possess such a device when they actually do not (Batesian mimicry). See also →coloration in marine animals.

mineral resources because of technologic and economic limits the mining of marine resources is mainly restricted to the continental shelves. Hydrocarbons (oil and gas) make up almost 90% of the current subsea mineral production. Other mineral resources of the continental shelves include heavy-mineral concentrates, gravel, shell, and lime mud mined by dredging in shallow water. Sulfur and salt are mined through drill holes. Phosphorite, potash, magnesium, fresh ground water, and geothermal energy (recoverable through drill holes) are other minerals that may be brought into production from the continental shelves in the future. The →manganese nodules that occur on the deep ocean floor are an enormous potential source of manganese, copper, nickel, cobalt, zinc and other metals. Exploration of the nodules, however, is not economically feasible yet because the mining of land resources is much cheaper.

mineralization the degradation of organic matter by organisms. During this process minerals (plant nutrients) become available again to the primary →producers. See also →decomposition.

Miocene see →geological time scale.

mitochondrion cell organelle of →eukaryotic organisms responsible for processes of respiration and electron-transport phosphorylation.

mixed layer water layer in which the vertical mixing is intense enough to give a homogeneous distribution of temperature and salinity as well as of other properties over its full depth. The term is mostly applied to the surface layer, where the →air-sea exchange causes strong convection and/or turbulence. It may be bounded underneath by a →pycnocline. Over large parts of the ocean the mixed layer varies seasonally and diurnally in depth.

mixed tides tides with an alternation of semidiurnal and diurnal tides, occurring in certain sea areas. See also →harmonic analysis of tides.

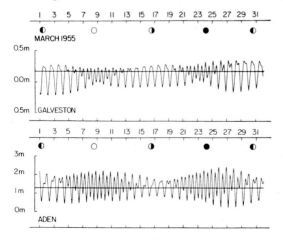

Mixed tides.
(After Scharnow
1978)

mixotroph organism able to assimilate organic compounds as carbon source while using inorganic compounds as energy source (as electron donor).

mode the most frequent value of a set of observations. See also →distribution.

models idealized representations of reality (the prototype). In oceanographic models the actual processes in the sea are simulated on the basis of underlying physical, chemical, and biological theories. They are used as a tool for further investigations and for predictions. Models are most advanced in physical oceanography but →ecological models are more and more used. Models are formulated in the form of a set of mathematical expressions (mathematical models) that are either solved analytically, or in numerical models are solved in a grid and with discret time steps. Furthermore there are analog models, especially hydraulic models, in which by some scaling of the governing equations the relevant processes are reproduced on a smaller scale and where their development is much faster than in reality. An essential problem in numerical models is the way in which the mathematical model is discretized. In hydraulic models the scaling requires that the relation between the relevant parameters, as expressed in nondimensional variables, remains the same (see →similarity theory). Models have to be calibrated by comparison with the observed past behavior of the prototype. In the →calibration certain parameters governing the behavior of the model can be adapted to obtain an optimal result. By sensitivity analysis the response of different elements in the model to external influences is investigated. The reliability of models for prediction is investigated by →validation, that is comparison with a set of independent observations of the prototype. Further development in modelling involves the regular incorporation of information from observations into the model (data assimilation) in order to improve model predictions.

Moho see →Mohorovičič discontinuity.

Mohole proposed deep bore hole through the earth's →crust to the →Mohorovičič discontinuity in order to obtain samples from the underlying mantle.

Mohorovičič discontinuity (abbr. Moho) boundary which separates the earth's →crust and mantle, characterized by an abrupt increase in seismic wave velocity and an increase in density from approximately 2.9 to 3.2 g cm^{-3}. It is situated 25 to 40 km below continents, 5 to 10 km below the ocean floor, and 50 to 60 km below mountain ranges. See also →earth's internal structure.

molds filamentous fungi. See →fungi.

Mollusca (mollusks) phylum (one of the largest) containing the snails, clams and all other bivalves, squids, octopods. There are some 50 000 living species and 35 000 fossil species reaching back as far as the →Cambrium. They

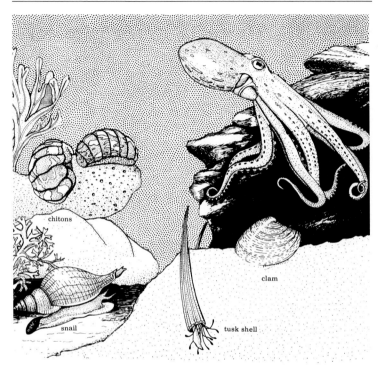

Mollusca. (After Keeton 1972)

occur in all marine biota, in freshwater, and on land. The unsegmented, soft-bodied animals are characterized by a muscular foot (the surface on which they crawl), a calcareous shell secreted by two lobes of the skin called the mantle, and often by the feeding organ in the mouth called →radula for scraping food. Between the mantle lobes and the body, a space called mantle cavity hides the "gills" or →ctenidia. The Mollusca are divided into the classes →Gastropoda (snails), the Monoplacophora, the →Polyplacophora (chitons), →Aplacopohra (solenogasters), the →Bivalvia, →Scaphopoda (tooth or tusk shells), and the →Cephalopoda (squids, cuttle-fish and octopods).

molting (syn. ecdysis) periodical renewal of the →exoskeleton in →Arthropoda, which allows these animals to change their body shape (during juvenile development), or to increase in size (growth). After an intermolt period, the cuticle loosens through the excretion of molting fluid from the hypodermis; is partly digested and softened, and thrown off. Then, the soft animal can change shape and/or increase in size, after which the new cuticle hardens.

momentum density momentum per surface area of a wave field. The momentum of the waves is found from the phase- and depth-averaged momentum that results from the →orbital motion. Because of the →Stokes drift there is an average momentum in the direction of the wave propagation. As a horizontal flux of momentum equals a →stress, the flux of wave momentum gives a so-called radiation stress, which can raise the water level when waves are moving into shallower water.

momentum equation equation in →hydrodynamics which formulates the equations of motion as a relation that expresses the conservation of momentum in a moving fluid subject to different forces.

monitoring the process of repetitive observation, for defined purposes, of one or more elements of the environment, according to prearranged schedules in space and time and using comparable methodologies for environmental sensing and data collection. Monitoring provides factual information concerning the present state and past trends in environmental behavior (this definition is based on the definition of monitoring accepted by the Governing Council of the United Nations Environmental Programme, UNEP). Biological monitoring has a public health objective (e.g., heart rhythm, activity of kidneys, lead concentration in human hair or finger nails). Both environmental and ecological monitoring are concerned with the natural environment (excluding human beings): its ecological, physical, and chemical characteristics. Environmental monitoring primarily deals with the recording of physical and chemical entities in the environment; sometimes the term is used for the registration of physical →stress on biota or–by using →indicators–the detection of trends induced by harmful substances. Exposure monitoring, the recording of total exposure to critical receptors, is therefore a member of the environmental monitoring family. Ecological monitoring is focused on the recording of changes in ecological entities to determine trends in →pollution or the extent of →contamination, geographical distribution, community composition etc. Two terms often used in connection with monitoring are →bioassay and →indicator.

monoclonal antibody is a monospecific antibody.

Monothalamia →Foraminifera with noncalcareous, single-chambered test. The pseudopodia are rarely reticulate and the protoplasma does not extend as a layer over the tests. Only some of the species are marine. Example: *Trichosphaerium*.

monsoon seasonal variation of the weather in the tropical Indian Ocean. Because of the heating during the summer of the Asiatic land mass to the north there is a variation in the pressure distribution that causes more or less a reversal of the normal trade wind pattern (see →climate). As a consequence, during the (northern) summer westerly winds prevail north of the equator, and in response the circulation pattern in the ocean shows marked seasonal variations.

morphology the study of form, e.g., of organisms, the sea floor, coastal forms, land forms (geomorphology).

mortality (syn. death rate) the number of individuals of a population dying in a given period, see also →population parameters. In fishery research, natural mortality (death caused by natural factors such as predation) is distinguished from fishery mortality (caused by catches by man).

mother of pearl see →nacreous layer.

mucoid feeder mechanisms mechanisms of collecting food by means of a structure made of mucus produced by the animal itself.

mucous glands generally unicellular glands which secrete mucin, a protein polysaccharide, which, with water, forms a lubricating solution called mucus.

mud slimy and sticky mixture of water and fine-grained particles with a consistency varying from that of a semifluid to that of a soft plastic sediment. Muds are generally described by color, e.g., black mud, gray mud, red mud.

mud flat see →tidal flat.

mudflow →sediment-gravity flow in which more than 50% of the solid fraction consists of particles smaller than 64 µm.

mutualism see →parasites.

Mysidacea (mysids) order of the class →Malacostraca, subphylum →Crustacea (phylum →Arthropoda). Small, →pelagic shrimp-like crustaceans

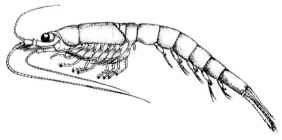

Mysidacea. (After Tattersall and Tattersall 1951)

living in freshwater, in the brackish environment, and in the sea. The thorax has a →carapace, which is not fused with the last four thoracic segments but covers them loosely. Another important characteristic of mysids is the →statocyst in the uropod endopodite (the base of the "tail"), which is visible as a small black dot from the outside. The first one or two pairs of thoracic appendages are →maxillipeds, the other six or seven are uniform and filamentous. Mysids live in large swarms along the seabed, and form an important food source for fish.

N

nacreous layer (syn. mother of pearl) the laminated inner layer of the calcareous part of the molluskan shell, secreted by the inner mantle surface. Loose particles between the inner shell surface and the mantle will generally be covered by nacre and may result in →pearls.

nan(n)ofossil see →microfossil.

nan(n)oplankton planktonic bacteria, algae and →Protista, between 2 and 20 µm in size. See →plankton.

Nansen bottle instrument for taking water samples from the water column named after the Norwegian explorer F. Nansen (1861–1930). The instrument, which is attached to a steel cable, is freely flushed while being lowered, and is closed at the given depth by a signal from the ship. Although other types have been devised, the original Nansen "bottle" consists of a tube that

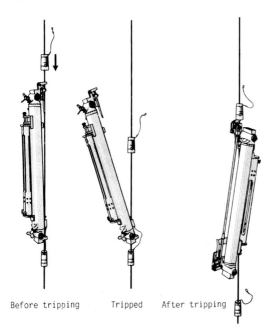

Before tripping   Tripped   After tripping

Nansen bottle.
(Neumann
and Pierson 1966)

is closed by a kind of cock on either side. The Nansen bottle is tripped by a messenger weight that is slipped along the cable, and the closing occurs when the upper one of the clamps by which the apparatus is attached to the cable loosens and the Nansen turns upside down. This may also fixate a →reversing thermometer that can be attached to the instrument, so that a water sample and a temperature observation are obtained simultaneously. It is possible to attach more than one Nansen bottle to a cable, and so obtain a series of samples in one so-called →cast.

natality (syn. birth rate) number of new born individuals entering a population per individual member per unit of time. See also →population parameters.

natural mortality see →mortality.

natural selection the key factor in Darwin's evolution theory. Survival of the fittest: depending on environmental conditions certain types of juveniles or adults have a lower →mortality than others with different genomes, so due to a larger reproduction the frequency of certain genes within the population will rise. New, favorable, mutants may change the species. See also →evolution.

nauplius development stage occurring in many →crustacean species. There are six naupliar stages (n1–n6). The larva is oval, unsegmented and has three pairs of appendages, which increase in complexity from n1 to n6. In →copepods, the nauplius changes into a →copepodite; in →Malacostraca into a →zoea.

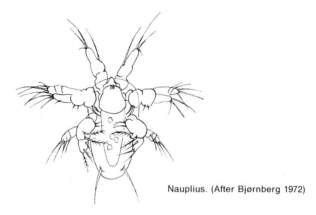

Nauplius. (After Bjørnberg 1972)

nautical mile (syn. sea mile) one-sixtieth part of a degree at the equator. The UK nautical mile is 1853.18 m, while the international nautical mile is 1852 m.

Navier-Stokes equation the equation of motion in its most elementary form. See →hydrodynamics.

navigation (biol.) see →orientation.

neap tide tidal period during which the effect of the semidiurnal →solar tide opposes the semidiurnal →lunar tide. The tidal range is small, the high water being lower and the low water being higher than normal. Neap tide occurs once every fortnight.

nektobenthos →benthic organisms living close to, but not on or in the sea bottom.

nekton all those →pelagic animals that are active swimmers such as most adult squids, fishes, and marine mammals, as opposed to →plankton, the passively drifting or weakly swimming organisms.

nematocyst thread cell, see →Cnidaria.

Nematoda (nematodes) phylum of the roundworms, which together with phyla like Gastrotricha and Rotifera and a few other phyla form the

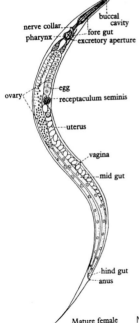

Mature female   Nematoda. (Borradaile and Potts 1959)

→Aschelminthes. Small, normally microscopic, cylindrical worms that live in large numbers (up to millions per square meter) in muddy bottoms in brackish and marine environments and in freshwater. They constitute an important part of the →meiofauna. Additionally there are free-swimming nematodes, and also terrestrial and parasitic forms. They have a mouth in front with palps, lobes, or bristles depending on their type of food or substrate, a smooth body with a well-developed three-layered skin or cuticle, and a digestive tract ending in an anus at the end of the body. They crawl or swim with undulating movements, eat algae, small animals, or decaying material. The sexes are generally separated, sometimes females attract males with a →pheromone, and after copulation eggs are deposited in clusters. There are →hermaphroditic forms; →parthenogenesis occurs as well.

Nemertea (an older name is Nemertini) phylum of mostly marine worms, also called Rhynchocoela. Elongated and often flattened predatory worms that generally live in the seabed, or under stones or seaweed. They resemble the →Aschelminthes, but are more elongated, while the anterior end of the body is normally pointed or shaped like a spatula. Typical of the phylum is a well-developed →proboscis, in front of the mouth, with which they grab or stab their invertebrate prey. Especially the larger species have the tendency to fragmentize when disturbed.

Nemertea. (Barnes 1987)

Nemertini older name for →Nemertea.

Neogene see →geological time scale.

nepheloid layer water layer on the shelf or in the deep ocean, usually near the bottom, where suspended matter concentrations are relatively high. On the shelf such a layer may be less than 10 to more than 20 m thick; in the deep sea its thickness may vary from 200 m to 1000 m.

nephelometer instrument for measuring the cloudiness (transparency) of water which is chiefly related to the concentration and particle size of the suspended material. Measures that are used for cloudiness are the light transmission (light attenuation) through the water or the scattering of light by the suspended particles.

nephridium excretory organ present in many →invertebrata and in →Cephalochordata, serves to convey the excretory and reproductive products from the →coelom to the exterior. Consists of a tube of →ectodermal origin opening to the exterior end. The other end may be closed with →flame cells opening into it; or may open into the coelom.

neritic originally defined by E. Haeckel (1834–1919) as the aquatic environment overlying the sublittoral zone (0 to 200 m), i.e., the water over the continental shelves, as a complement to the oceanic environment of the waters over the deep sea. Paleontologists have also applied the word "neritic" to the environment of the bottom itself.

net primary production see →primary production.

nets fishing gear of a wide variety of designs and meshed materials, developed by man since prehistoric times to collect fish and other food animals. A division can be made into passive nets, which are vertical screens of netting floating in the water (as, e.g., in the drift-net fisheries for herring) or attached to poles standing in shallow water (gill nets). The main idea is that

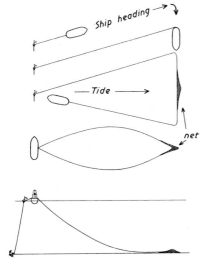

Nets. (Wimpenny 1953)

migrating fish swim through the meshes with the head, and cannot move back because of the gill covers. Other passive nets are fykes and different types of fish traps, in which migrating fish are trapped. In marine science, nets for the collection of organisms from water or seabed show developments based on the same principles. Plankton is collected with fine-meshed nets (→mesh size as small as 50 or even 20 μm), mostly conical in shape with a steel ring in the opening. For the collection of the larger →zooplankton, such nets are mounted inside funnel-shaped metal bodies which can be towed at speeds of several →knots. When the catch in such a torpedo is continuously enveloped and preserved during sampling, the gear is called →Continuous Plankton Recorder. Different depth layers can be sampled with the Multi-net, a rosette of small plankton nets with opening and closing mechanisms towed vertically upwards. For the collection of macroplankton (e.g., jellyfish, mysids, euphausids, and fish larvae) pelagic plankton trawls are available, such as the Isaacs-Kidd Midwater Trawl and RMT 1 + 8. These large nets with mesh sizes of 1 to 5 mm are generally towed from the A-frame of research ships, while depressors are mounted to force the nets downwards; →acoustic releases may be used to open and close these nets under water in order to sample a certain depth interval. The bongo net is special in that it actually consists of two conical plankton nets with circular openings attached symmetrically to both sides of the fishing line, to reduce the disturbing effect of lines and connections in front of the net opening.

neuston the typical community of bacteria, plants and animals living at the very surface. Species living day and night in or above the microlayer form the euneuston (including the only marine insect *Halobates*); species not attached or confined to the surface form the pseudoneuston; buoyant organisms subject to wind drift form the pleuston (*Sargassum* weed, the coelenterates *Physalia, Velella*). See also →surface film.

neutral surface see →isentropic surface.

new production primary production based on newly incorporated nutrients, imported into the →euphotic zone from elsewhere, as opposed to regenerated production based on nutrients recycled in the euphotic zone. The rate of export of organic matter from the euphotic zone by either sedimentation or transport with the currents, and new production by →phytoplankton may be unbalanced over short periods but they have to be in balance over a 1-year period if phytoplankton is to maintain itself.

niche the role or function of a species in the community or →ecosystem (often wrongly used for the →habitat in which a species occurs).

Niño see El Niño.

Niskin bottle see →seawater sampler.

nitrate inorganic nitrogen compound in which the nitrogen atom is in its highest oxidation state. Chemical symbol NO_3^-. In nature nitrates are the end product of →aerobic degradation of nitrogeneous organic compounds. They are an important nitrogen source for →phytoplankton growth, especially when ambient →ammonia concentration is low. In →anaerobic systems nitrates can act as electron acceptors for the oxidation of organic compounds (→denitrification). In this microbiological process nitrogen gas (N_2) is the main end product.

nitrification the process in which →ammonia is oxidized by bacteria to nitrate, see also →nitrogen cycle.

nitrogen (N) the most abundant element in the atmosphere of the earth, as N_2 gas (ca. 78 vol%). Physical constants: mol. weight N_2 28.016; density 1.2506 g dm^{-3} (gas), 0.885 (liquid); melting point $-210\,°C$; boiling point $-195\,°C$. Various forms of N compounds are present on earth, reduced forms as NH_3, N_2H_4, N_3H, and oxidized forms as N_2O, NO, N_2O_3, N_2O_4, and N_2O_5. Nitrogen is also an essential element in all living organisms. As building atom in amino acids it is one of the main compounds of proteins. Some organisms (some Bacteria and Cyanobacteria) can fix atmospheric nitrogen. In this way they fertilize (eutrophicate) soils and waters with nitrogen compounds. See also →nitrogen cycle.

nitrogen compounds inorganic or organic molecules which contain one or more nitrogen atoms in their structure. Most naturally occurring inorganic nitrogen compounds in the atmosphere are nitrogen gas (N_2), with a relative abundance of 78%, and the gaseous nitrogen oxides N_2O, NO, NO_2. The nitrogen oxides are partly produced by burning of fossil fuel and are involved in the →greenhouse effect as well as in the stratospheric ozone destruction. Another inorganic atmospheric nitrogen compound is →ammonia (NH_3), which indirectly contributes to the "acid-rain" problem. Inorganic, dissolved forms of nitrogen compounds are →nitrates (NO_3^-), and ammonium (NH_4^+), which can both be nitrogen sources for →phytoplankton growth and thus be responsible for the environmental problem of →eutrophication. Important small organic nitrogen compounds which can be found in nature as a consequence of anabolic or catabolic (see →metabolism) processes are urea, amines and →amino acids. Nitrogen atoms are present in all organisms (→proteins, nucleic acids).

nitrogen cycle →biogeochemical cycle executed by various kinds of organisms in which nitrogen is involved. In its most inert form, nitrogen gas makes up most of the atmosphere. In living organisms, N is a major constituent of →amino acids, proteins, nucleic acids. However, despite the enormous amount of →nitrogen fixation which takes place, only a few highly specialized →Bacteria and many →Cyanobacteria are able to extract nitrogen from air and turn it into biologically available compounds.

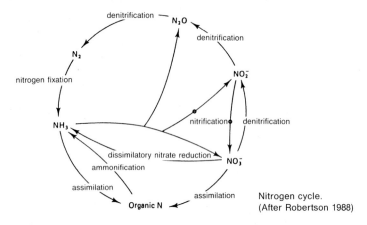

Nitrogen cycle.
(After Robertson 1988)

Nitrogen gas is also fixed by chemical reactions in the atmosphere and by industries. Conversely, nitrogen is lost to the atmosphere by activities of Bacteria which reduce oxidized nitrogen compounds to nitrogen gas and various other volatile compounds (→denitrification). Between these two extremes, a host of reactions take place including the oxidation of →ammonia to nitrite and →nitrate (→nitrification), the reduction of these oxidized nitrogen compounds to ammonia (dissimilatory nitrate reduction) and the incorporation of the various compounds in organic molecules (N-assimilation). The release of ammonia from nitrogenous organic material during degradation is called ammonification.

nitrogen fixation the reduction of nitrogen gas to →ammonia. This fixation of molecular nitrogen is carried out by various →Bacteria and →Cyano-bacteria in water as well as in soil. Nitrogen-fixing plants have tumor-like structures (nodules) produced by their roots and containing nitrogen-fixing bacteria. This is a →symbiotic relation. For the production of fertilizers atmospheric nitrogen is fixed on an industrial scale. Its large-scale use in agriculture can result in an excess of nitrogen compounds in →ecosystems and thus promote the formation of algal →blooms.

noble gases nonreactive elements of group zero of the periodic system. In seawater helium, neon, argon, krypton and xenon are detectable and occur at levels of: $4 \ 10^{-5}$, $1.5 \ 10^{-4}$, $3 \ 10^{-1}$, $6 \ 10^{-5}$ and $7 \ 10^{-6}$ ml kg^{-1}, respectively, calculated on the basis of their apparent solubility.

Noctiluca a large, motile, →phagotrophic →dinoflagellate named after its ability to emit light. See also →bioluminescence.

node in a →standing wave there are points called nodes where the phase reverses and the amplitude is zero. In oscillating seas at nodes, the sea level is constant and the alternating current is maximal. In astronomy a node is the crossing-over point of the equator and the track of a celestial body (such as the moon). The nodes of the moon gradually move along the equator and this is the cause of a variation in the tidal amplitude with a period of about 18 years (nodal factor).

nondimensional parameters see →similarity theory.

notochord skeletal rod of large vacuolated cells packed within a firm sheath. Present during any stage of development in all →Chordata.

Nuda class of the phylum →Ctenophora, without tentacles.

nursery area an area which offers favorable circumstances for food (growth) and shelter for juvenile stages of the population of a certain species.

nutrients term that refers to inorganic compounds which are essential for →phytoplankton growth. They consist of N-compounds like →ammonia, →nitrate, and nitrite, P-compounds like HPO_4^{2-} and $H_2PO_4^-$, which are taken up by algal cells and are built in as atoms in →amino acids, →proteins, nucleic acids, fats, etc. Also dissolved silicate has to be considered a nutrient because it is essential for diatoms to build up their skeletons, which consist of biogenic silicate (→opal). The inorganic N-, P-, and Si-compounds are often referred to as macronutrients to discriminate them from metal ions, the so-called →micronutrients, which are essential because they are part of enzyme systems, but are taken up by the cells in much smaller quantities than the macronutrients. Because uptake of nutrients by algae is restricted to the upper part of the water column, the concentrations of these compounds there are usually depleted compared to the deeper part of the water column where dissimilation processes dominate over assimilation returning nutrients to the water column.

nutrient uptake see →Michaelis Menten kinetics.

Nyquist frequency see →aliasing.

O

ocean originally the great stream or river supposed to encompass the disk of the earth and personified in Greek mythology by Oceanus, the God of the great primeval water, the son of Uranus and Gaia, and husband of Tethys. Hence the great outer sea as opposed to the →Mediterranean. Now the vast body of water on the surface of the globe, which surrounds the land, covering ca. 70 % of the earth's surface.

ocean-atmosphere exchange see →air-sea exchange.

ocean dumping the marine environment is by many countries considered a disposal region for →waste products such as waste effluents, solid wastes from sewage →sludges, dredge sludges, and radioactive and chemical wastes. Since the birth of the Atomic Age, nuclear nations have deposited part of their unwanted radioactive waste in the sea. The earliest sea-dumping program was started in 1946 by the US and was terminated in 1960. European nations have dumped at sea for many years, particularly at one site in the Atlantic Ocean, the NEA (Nuclear Energy Agency) dumpsite 400 miles off the coasts of England and Portugal. Under pressure of public opinion most countries have stopped sea dumping. Research on the NEA dumpsite continues to measure possible effects on →benthic organisms. The feasibility of deep burial of high-level radioactive wastes in ocean sediment has so far not resulted in possibilities that would guarantee the safeguarding of the marine environment. Ocean dumping falls under the restrictions of the Oslo Convention.

ocean dynamics the application of →hydrodynamics (fluid mechanics) to the ocean, where specific problems of large-scale horizontal motion over a rotating earth play an important role. As in ocean dynamics, the reference framework for the equation of motion is usually a fixed point that rotates with the earth, a fictitious force has to be introduced: the →Coriolis force. Furthermore in most cases the vertical motions are small enough to be neglected, so that in the vertical direction the hydrostatic pressure equals the weight of the overlying water per surface unit, as expressed in the hydrostatic equation. The general →Navier-Stokes equation of →hydrodynamics is often not practical for solving oceanographic problems, but they can be reduced to approximate equations applicable to special situations. In the ocean the variations that occur in water density are often neglected insofar as this does not concern the pressure distribution via the hydrostatic equation. This approach is called the →Boussinesq-approximation. Another usual approximation is only to consider the large-scale motions and to incorporate the small-scale motion in →turbulence.

ocean exploration precedes →oceanography. Primarily exploration at sea was aiming at the investigation of new shipping routes or at finding new land. A transition to real ocean exploration took place with the voyages of James Cook (1728–1779), one of the greatest explorers and navigators. He made three round-the-world voyages over a period of 12 years, during which he destroyed the myth of the southern continent. He claimed Australia and New Zealand for the British Crown, charted the islands of the Pacific. Used the *Endeavour* on his first voyage (1768–1771) on which he was accompanied by the naturalists Banks and Solander. He was murdered on Hawaii. Later explorations often involved scientists (e.g., Darwin on the voyage of the Beagle (1831–1836) with captain R. Fitzroy, see also →evolution). The →Challenger Deep Sea Expedition (1872–1876) marked the beginning of oceanographic →expeditions.

oceanic basement oceanic crust below sedimentary deposits, extending downward to the →Mohorovičič discontinuity. The top of the basement usually forms a clear →reflecting horizon in seismic →reflection shooting. See also →layer 1, layer 2, etc.

oceanic crust see →crust.

oceanographic station geographic position at sea where a research vessel is stopped to allow measurements and sampling.

oceanography literal meaning: description of the oceans. Used for the scientific studies of ocean, its boundaries and bottom topography, its physics and chemistry and of its marine organisms, including the interrelations and interactions. Sometimes the word "oceanology" is used instead, but this may also be used to incorporate marine technology. Traditionally, the →Challenger Deep Sea Expedition (1872–1876) is taken as the starting point of modern oceanology.

Octopoda (octopods) order of the class →Cephalopoda (phylum →Mollusca). Octopods are cephalopods with eight suckered arms and a globular body, a highly developed pair of eyes and a horny bill-like mouth. e.g., *Octopus vulgaris*, living along rocky coasts of southern Europe.

oesophagus see →esophagus.

offshore region starting from →baseline seawards. See also →coastline.

O-group a term used in connection with year classes, to refer to the animals in their first year of life from the end of the egg stage onwards.

oil spills unfortunate but regular phenomena in the marine environment, due to the release of oil from ships during tanker or platform accidents, or due to the discharge of effluents during the cleaning of oil tanks. Oil spillage control is regulated by IMO (International Maritime Organization, London). Oil spills are particularly hazardous to sea birds, since oil contamination destroys the natural protective grease system of their feathers so that they get wet and die of cold. Oil spills can be treated with →detergents, which disperse and solubilize the →hydrocarbons in the water. This also facilitates their breakdown by bacterial decomposition.

olfaction see →homing.

oligo- prefix meaning few.

Oligocene see →geological time scale regions.

Oligochaeta (oligochaetes) class of the phylum →Annelida or segmented worms. These are the annelids with few or no →chaetae or bristles on the segments, in contrast to the class →Polychaeta. The class includes terrestrial forms like the common earthworms, freshwater species, and ca. 200 marine species among a total of ca. 3000. They are mainly →deposit feeders and can reach high densities in shallow estuarine →mud flats. They are all →hermaphrodites; the larval development occurs in the yolk-containing eggs. Example: *Enchytraeus, Tubifex.*

oligotrophic see →eutrophic regions.

omnivore omnivorous organism, animal that devours or feeds on all kinds of food as opposed to specialized feeders like →carnivores, which eat only other animals or →herbivores, which feed only on plants.

ooze →pelagic sediment consisting of ≥ 30% skeletal remains of pelagic organisms (either calcareous or siliceous), the rest being →clay minerals. Oozes are named after (the chemical composition of) their characteristic organisms, e.g., foraminiferal ooze, diatom ooze, but also calcareous ooze and siliceous ooze.

opal amorphous silica, SiO_2 with 5 to 20% of$ H_2O. In marine sediments opal-A (biogenous silica) occurs in the form of the →frustules of →diatoms and the skeletons of →radiolarians, silicoflagellates, choanoflagellates and sponges, especially around Antarctica and in equatorial and upwelling regions. It is not stable and is subject to dissolution. Saturation of pore waters with silicic acid and some impurities, notably Al, retard dissolution. On time scales of tens of millions of years, opal-A is transformed via opal-CT into quartz. Near →hydrothermal vents hydrothermal opal, which is less pure than opal-A, may be precipitated.

operculum (1) cover of gill-slits of fish; (2) exoskeletal plate of →Gastropoda, which can close the shell opening when the animal withdraws into the shell; (3) present in Bryozoa, see →Cheilostomata.

Ophiuroidea subclass of the class →Stelleroidea (phylum →Echinodermata). Brittle stars or serpent stars. Starfish with a relatively small central disk (up to ca. 3 cm) with long and flexible arms that are markedly separated from the disk. Other differences with the starfish subclass →Asteroidea are the absence of →ambulacral groove and the fact that the ambulacral feet or

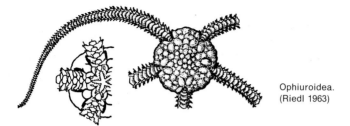

Ophiuroidea.
(Riedl 1963)

podia play no role in locomotion. The Ophiuroidea have many species and are found in all types of marine habitats, especially in mud bottoms, where they live with the disk buried, and the sometimes branched arms protruding from the seabed for the collection of particulate food. The typical serpent stars belong to the Ophiurae with examples: *Ophiura* (family Ophiolepididae) and *Amphiura* (family Amphiuroidea).

Opisthobranchia subclass of the class →Gastropoda (phylum →Mollusca), which have undergone torsion, but show a reversal of torsion (detorsion). The →ctenidia are posterior to the heart, the mantle cavity opens laterally or posteriorly, and the visceral nerve loop is not twisted into the figure of an

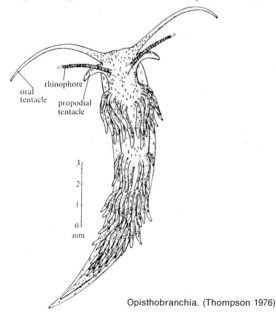

rhinophore
oral
tentacle
propodial
tentacle

3
2
1
0
.mm

Opisthobranchia. (Thompson 1976)

eight. The ctenidia and mantle cavity are sometimes absent when respiration is through the body surface, which often develops secondary gills. The shell is usually reduced or absent. They are →hermaphrodites.

optics in the sea the optical properties of water influence the penetration of sunlight, which is an important factor in →primary production. Furthermore, optical properties give information on the distribution of →water masses and on the content of plankton and non-biogenic particles for instance by means of →remote sensing. The optical properties depend on the wavelength. They are: (a) the absorption of light by the water and by dissolved organic substances (such as →Gelbstoff) or by absorbing particles; (b) the scattering of light (by the water molecules and by small suspended particles) and (c) →refraction. Absorption and scattering both decrease the amount of energy in a beam of light, and together they give the attenuation (sometimes indicated with extinction) of light. Temperature and salinity influence the refraction, but the absorption is practically independent of temperature and salinity. Light falling on the sea surface is partly reflected (see →albedo). In the water the light is attenuated by absorption and scattering. To obtain the rate of attenuation two quantities can be considered: the →irradiance and the →radiance.

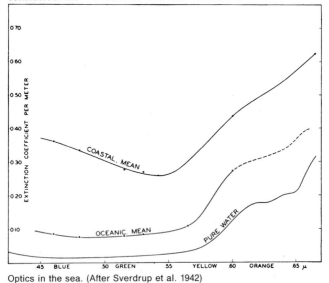

Optics in the sea. (After Sverdrup et al. 1942)

orbital motion in a →short wave the individual water particles describe a (in first approximation) circular path in the vertical plane, with a radius that

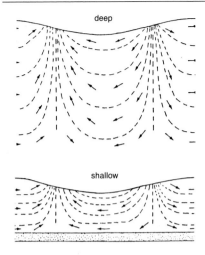

Orbital motion.

decreases exponentially with depth. At the surface the radius is equal to the wave amplitude, and the orbital speed is the circumference of the path divided by the wave period. If the wavelength becomes larger the depth of the water cannot be neglected (→long wave) and the circular path becomes gradually elliptical, becoming more and more elongated toward the bottom. In a higher-order approach the orbital paths are not completely closed after one revolution, giving a net water transport in the direction of wave propagation (see →Stokes drift).

orbital variations the amount of solar radiation striking the earth's upper atmosphere at any given latitude and season is defined by three components of the earth's orbit around the sun: the excentricity (periods of 413 000, 123 000 and 95 000 year periods), obliquity tilt or inclination of the earth's axis (41 000 and 39 000 y), and precession or movement of the vernal point (23 000 and 19 000 y). The geometries of past and future orbits are known as the →Milankovitch orbital variations. These variations are now considered major forcing functions in the glacial to interglacial oscillations that marked the →Quaternary period. See also →light. (Figure see p. 190)

Ordovician see →geological time scale.

organic carbon see carbon, organic.

organotrophs see →metabolic diversity.

orientation of organisms (static) is controlled by environmental factors. It can be modified by changes in growth direction in attached organisms (→benthic algae, corals) and by movements in free-living organisms. (Dynamic) orientation in space requires series of stimuli for directional

ICE GROWTH ON NORTHERN HEMISPHERE

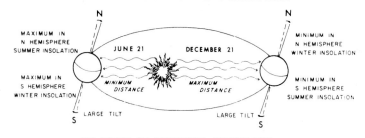

ICE DECAY ON NORTHERN HEMISPHERE

Orbital variation. (After Kennett 1982)

control of the movement and for the initiation of the movement. Visual, auditory, tactile, and chemical senses can be used. Many organisms keep compass course by using celestial cues. Other cues may be light, temperature, salinity, organic substances, substrate, pressure, sound, gravity, electrical and magnetic fields, etc. See also →homing, migration, invasions, magnetotaxis.

Ortman line dividing line in the northern and soutern hemispheres between the warm-temperate and cool-temperate zones; a zoogeographical term often used in taxonomic literature.

osmoregulation the capability of an organism to control the composition of its extracellular fluids independent of the osmotic concentration in the external environment. The significance of a stable extracellular osmotic concentration is that it prevents the body cells from shrinking (by loss of water) when external osmotic concentrations increase, or, from swelling (by water inflow) at decreasing external osmotic concentrations. All vertebrates, some marine invertebrates, most brackish-water invertebrates, and all freshwater species show some degree of osmoregulation.

Osteichthyes see →Pisces.

Ostracoda class of the subphylum →Crustacea (phylum →Arthropoda).
Mussel-shrimps. Small (a few mm) shrimp-like crustaceans, with a calcified
→carapace in two valves completely enclosing the body. They occur in
freshwater and all marine habitats, →benthic as well as free-swimming. The
food varies, depending on the species, from detritus and plant material to
even small fish. There is also a wide variety of fossil species, restricted to
discrete depth ranges, which makes them useful depth indicators, while their
rapidly evolving lineages make them useful as stratigraphic markers (i.e.,
tools for correlation). See →biostratigraphy.

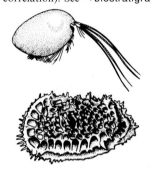

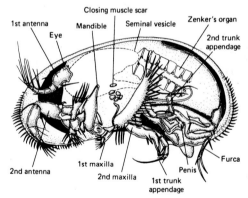

Ostracoda. (From Barnes
1987 after Cohen 1982)

otolith ear-stone of fish, which generally shows distinct daily and annual
growth rings, which are a measure for the age of the individual. Used as
→statoliths (see →statocysts). The growth rings can be used in →age
determination of the fish. Otoliths are species specific, and are usually the
only parts of the fish that fossilise.

otter trawl see →trawl.

outwelling a process of export of nutrients and organic matter from an estuary into the ocean, enriching (fertilizing) the ocean adjacent to the →estuary.

ovary organ which produces eggs. In vertebrates it also produces sexual hormones.

overflow continuous or intermittent flow of relatively heavy water from one sea area to another over submarine ridges separating them. Overflow occurs over the sill in the Strait of Gibraltar, the Bab-el-Mandeb and the ridges between Scotland, Iceland, and Greenland.

overgrazing see →grazing.

oviparous mechanism of reproduction in which the offspring develop in eggs that are deposited and hatch outside the body of the female animal.

ovoviviparous mechanism of reproduction in which the offspring develop in eggs inside the female body, and leaves the female animal as juveniles.

oxygen (O) the most abundant element in the earth's crust (46.4%), followed by →silicon (28.2%) and →aluminum (8.3%). After nitrogen the most abundant gas (O_2) in the atmosphere of the earth (20.95%). Physical constants: mol. weight 31.9988; density 1.429 g dm^{-3}(gas 0 °C), 1.149 kg dm^{-3}(liq. -183 °C); melting point -218.4 °C, boiling point -182.962 °C. Ozone (O_3) is formed by dis- and reassociation of oxygen with UV light (<240 nm) at heights of 30 to 50 km in concentrations of 20 ppm. The ozone layer effectively shields the lower parts of the atmosphere from the destructive effects of short wave UV light on living organisms. Space research has confirmed older spectroscopic evidence of very low free oxygen concentrations in the atmosphere of other planets in the solar system. Under purely physical conditions low steady-state oxygen concentrations are formed by photochemical dissociation of water vapor and geological evidence points to an almost anoxic primordial atmosphere of the earth. The evolution of life allowed the gradual formation of an oxic atmosphere and a hundredfold increase of oxygen pressure to the present level possibly took place in the last few hundred million years with the development of plants capable of →photosynthesis. The hydrosphere contains dissolved oxygen in about 1% of the quantity present in the atmosphere.

oxygen demand the consumption of oxygen in incubated samples at a certain constant temperature during a prescribed timespan. BOD (→biological oxygen demand) values are frequently used as water quality indicators in the analysis of natural waters. See also →chemical oxygen demand (COD).

oxygen isotope ratios on earth the chemical element O consists mainly of isotope ^{16}O ($\pm 99.758\%$), with minor amounts of isotopes ^{17}O ($\pm 0.038\%$) and ^{18}O ($\pm 0.204\%$). The H_2O in seawater contains O in roughly the same isotopic proportions. Minor variations in the isotopic ratio $^{18}O/^{16}O$ can largely be ascribed to temperature effects. Another cause of the different ratios is the slight difference in vapor pressure, causing higher $^{18}O/^{16}O$ ratios in surface waters with high evaporation, and lower ratios in deeper water. See also →isotopes.

oxygen minimum layer water layer in the oceans below the →euphotic zone, where, due to settling of organic →detritus and the presence of vertically migrating zooplankton during the day, bacterial oxygen consumption and respiration are relatively high.

oxygenic photosynthesis see →photosynthesis.

P

paleo- (or palaeo-) prefix meaning old, ancient, remote in the past.

paleobotany the study of the plant life of the geologic past.

Paleocene see →geological time scale.

paleoclimate climate in the geologic past.

paleocurrent ancient current that existed in the geologic past.

paleodepth the depth in the past.

paleoecology the study of the interrelationships between ancient organisms and their environment.

Paleogene see →geological time scale.

paleontology the study of fossils, see also →micropaleontology.

Paleozoic see →geological time scale.

pancake ice see →ice.

Pangea the early super continent (from late Carboniferous to Trias, see →geological time scale) from which present-day continents have originated through fragmentation and →continental displacement. This supercontinent was surrounded by the super ocean →Panthalassa and broke up into →Laurasia to the north and →Gondwana to the south during Jurassic times. (Figure see p. 195)

Panthalassa the →superocean surrounding →Pangea.

paralytic shellfish poisoning (PSP) a serious illness caused by eating →shellfish, which themselves have consumed toxin-producing microscopic organisms, usually the →dinoflagellate *Gonyaulax catenella*. The toxin is a potent nerve poison; early symptoms of PSP are a tingling of the lips and tongue within minutes of eating. Depending on the amount of toxin consumed, symptoms may progress to tingling of fingers and toes, loss of control of arms and legs, followed by difficulty in breathing. Death can result (in approximately 15% of reported cases), due to paralysis of the breathing mechanisms in as little as 2 h. See also →diarrhetic shellfish poisoning.

parametrization a method to reduce mathematical expressions describing a complex process to simplified expressions that are easier to solve or more open to experimental verification. Parameters formulating the effects of less accessible elements of the process are introduced in a simplified equation.

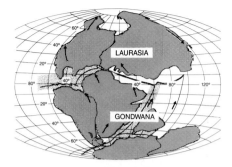

180 million years ago

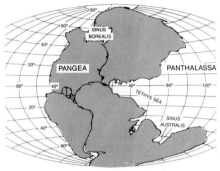

225 million years ago

Pangea. (After Dietz
and Holden 1970)

The parameters have to be chosen in such a way that the solutions of the
reduced equations describe the process with sufficient accuracy. An example
is the replacement of the →Reynolds stress, consisting of the cross-product
of small-scale fluctuations, by the product of an →eddy viscosity coefficient
and a gradient of the mean value.

parasites organisms that live inside (endoparasites) or attached to (ectoparasites) an individual of another species: the host, where they consume body
tissue and/or fluids without killing the host. Some species are facultative,
other obligate parasites (i.e., incapable of surviving without a host).
Parasitic species can have very complicated life cycles and more than one
host species may be involved. Kleptoparasites are not parasites in the strict
sense, but take the food collected by other species. See also →symbiosis.

parenchyma (1) in plants: tissue of thin-walled cells, not elongated in any
direction. (2) in animals: loose tissue consisting of irregularly shaped

vacuolated cells, forming a large part of the body of, e.g., the →Platyhel-minthes.

parthenogenesis spontaneous development of an egg (without fertilization) into a new individual; genetically identical to the mother.

partial pressure is the gas pressure exerted by a designated component of a gaseous mixture.

partial tide the astronomical theory of →tides states that tides contain many components of different periodicity: the partial tides, tidal components, or tidal constituents. In the →harmonic analysis of tides these different components are actually analysed from a tidal record. The partial tides are indicated with letters indicating the astronomical origin of this component (e.g., M = moon, S = sun) and a number or letter indicating its periodicity (1 = diurnal, 2 = semidiurnal, f = fortnightly, m = monthly, sa = semiannual, a = annual).

particle counter device automatically counting small particles in gaseous or aqueous media, often with electro-magnetic, electric or optical sensors. An electric type was originally developed by Coulter Electronics to automatize blood-cell counting in medical laboratories (→Coulter counter); its principle is based on changes in electrical conductivity induced by particles passing through a small opening. Particle counters are widely applied in industrial and aquatic research, e.g., in marine biology to measure the concentration and the size or volume of particulates in the sea. Modern particle counters discriminate particles from 0.3 to over 1000 µm in size, at a rate of up to 1000 per second.

particulate organic carbon (POC) all →organic carbon contained in particulate organic matter. Usually separated from →dissolved organic carbon (DOC) by filtering over 0.45 µm filters. Contained in living →phytoplankton, →zooplankton, bacteria as well as in dead organic particles (→detritus). In fact a continuum of organic matter occurs in the sea from small organic molecules to the largest particles (whales), the separation of DOC and POC by filtering being artificial and resulting in the inclusion of colloids, small bacteria, and other particles, both living and dead, in the dissolved fraction.

particulate organic matter (POM) the fraction of total organic matter that can be filtered or centrifuged from seawater. Hence the difference between →particulate and →dissolved organic matter (DOM) is arbitrary. See also →particulate organic carbon.

passive continental margin see →continental margin.

passive drift see →drift.

patchiness clustered occurrence of organisms. Causes are nonrandom distribution of physical and chemical properties of the environment, or certain behavioral patterns in organisms. Patchiness is one of the main difficulties in the interpretation of quantitative samples.

pathogen organism, e.g., a →parasite, bacterium or →virus, which causes disease.

P/B ratio see →turnover rate.

pearls hard structures of →calcium carbonate and →protein produced around an irritant inside the →nacreous shells of →mollusks such as the pearl oyster (*Pinctada margaritifera*). Cultivated pearls are produced in captured oysters by inserting a graft into the gonads. Pearl-like structures produced by nonnacreous mollusks such as the queen conch (*Strombus gigas*) or the giant clam (*Tridacna maxima*) are not considered to be true pearls.

pelagic the aquatic environment in deeper fresh or marine waters, of which the bottom and the →benthic layer make no part. In purely pelagic circumstances, there are only vertical gradients, like those of light, temperature, and salinity. These three gradients together with gravity may be the only cues for orientation of animals and plants. This means that life in the pelagic environment requires special adaptations such as those to prevent sinking. The term pelagic is widely used to indicate that certain organisms are adapted to live in the pelagic: pelagic plants are the planktonic algae, pelagic animals are micro-, meso- and macroplankton; pelagic fish are those living in the water column. Various depth zones can be discerned in the pelagic environment (see →zonation). Usually marine →ecosystems are divided into two different compartments: the benthic system at the bottom and the pelagic system in the overlying water column.

pelagic sediment see →deep-sea sediments.

pelagic trawl see →trawl.

pennate diatoms see →Bacillariophyceae.

peptide compound formed of two or more (polypeptide) →amino acids, by an amino (NH_2) group of one joining with a carboxyl (COOH) group of the next, forming peptide links (-NH-CO-) with elimination of water.

Peracarida superorder of the class →Malacostraca, subphylum →Crustacea (phylum →Arthropoda). Contains the seven orders that have in common that the females have a ventral brood-pouch or marsupium: →Mysidacea, →Cumacea, →Tanaidacea, →Isopoda, →Amphipoda and the small orders Thermobaenacea (eight species) and Spelaeogriphacea (one species only).

pericardial cavity cavity in which the heart lies. In →Arthropoda and →Mollusca: a hemocoelic space supplying blood to the heart.

Peridinia see →Dinophyceae.

periodic table table of chemical elements published in 1869 by D.I. Mendelejeff (1834–1907), schematizing all chemical elements according to their atomic number and atomic weight, taking into account their valencies and general chemical properties. In marine chemistry the table is of the utmost importance for a good understanding of the physical/chemical properties of elements in the oceans.

periodicity see →biological clock.

periphyton (syn. Aufwuchs) miscellaneous assemblage of organisms (both plants and animals) attached or clinging to all free surfaces of objects submerged in water (plants, wood, stones, rocks, etc.); best developed in the →sublittoral zone.

perivisceral cavity see →coelom.

Permian see →geological time scale.

pesticide(s) (1) any substance or mixture of substances intended for preventing, destroying, repelling, or mitigating any pest (insect, rodent, nematode, fungus, weed, other forms of terrestrial or aquatic plant or animal life or viruses, bacteria, or other microorganisms) on or in living man or animals; (2) any substance or mixture of substances intended for use as a plant regulator, defoliant or desiccant. Pesticides can be divided into a number of groups of chemicals: organic pest control, organometallic pest control, inorganic agricultural, environmental pest control, degradation products and metabolites, drugs used in agriculture and food production and other materials used in agriculture and food processing, such as antioxidants, surfactants, dispersants, spreaders, thickeners, solvents, propellants, etc. More than 2000 different chemicals are in use as pesticides. Depending on the purpose of use, pesticides may have a great variety of functional groups.

pH a measure of the concentration of →hydrogen ions. pH is defined by $pH = -\log[H^+]$. A pH value of 7 is neutral. The pH range 0 to 7 is called acidic and the range 7 to 14 alkaline. The pH of seawater is relatively constant and seldom outside the range of 7.8 to 8.3 in the →photic zone of the open sea. This small range is due to the buffering action of the carbonic acid system. High pH values are observed at the end of vigorous →phytoplankton blooms, when considerable amounts of free carbon dioxide have been converted to organic matter by →photosynthesis. The opposite occurs in winter in temperate seas. In deeper layers, where photosynthesis is absent and organic matter oxidation is a continuous process, the pH is always low. See also →alkalinity and →ionic product of water.

Phaeocystis see →Haptophyceae and →sea foam.

Phaeophyceae (brown algae) class of the division →Heterokontophyta, multicellular algae ranging from microscopic plants to kelps of more than

50 m long. They contain →chlorophyll-c and fucoxanthin, and store dissolved laminarin as the chief carbohydrate reserve. Reproductive cells are formed in special parts of the plant-body, each cell liberating one →zooid. The genus *Fucus* and relatives produce eggs. Most brown algae prefer cold water. They all start their lives attached to rocks or other substrates but some survive detachment. The drifting *Sargassum* of the Sargasso Sea harbors a special life community. Kelps are harvested for their cell wall material, →alginate. *Fucus* is used as soil fertilizer in some coastal regions. Few fossils can be assigned unequivocally to Phaeophyceae.

phagocytosis is the ingestion of particulate material such as bacteria by protozoa and phagocytic cells of higher organisms. The cell simply engulfs the food particle.

phagotroph organism which feeds by the process of →phagocytosis. In the process the phagotroph engulfs a food particle from its surrounding into its cytoplasm, e.g., in a membrane-covered vacuole, where digestion takes place.

pharynx part of the alimentary canal between the oral (mouth) cavity and the →esophagus.

phase speed see →wave characteristics.

pheromones organic substances released by the exocrine glands of some organisms to the environment, in order to influence the behavior of other members of the same species. Pheromones thus represent a means of communication and transferring information by smell or taste. They evoke specific behavior, developmental or reproductive responses in the recipient, which may be of great significance for the survival of the species.

phoresis see →symbiosis.

phosphorescence physical phenomenon denoting the re-emittance of light absorbed by certain materials for several seconds, or even several minutes, after the excitation radiation has stopped. There is no essential difference between phosphorescence and fluorescence, except that the former persists after the excitation has ceased, whereas in the latter the light is re-emitted within 10^{-7} s. The light produced by some marine organisms (→bioluminescence) is the result of chemiluminescent reactions (in which a substrate is oxidized) and should not be confused with phosphorescence.

phosphorites marine deposits consisting largely of minerals dominated by the occurrence of the element phosphorus.

phosphorus cycle phosphorus (P) occurs in seawater in living organisms (LOP) or as dissolved inorganic P (DIP), dissolved organic P (DOP) and particulate P (POP). In surface waters LOP is usually highest and DIP low, due to uptake during primary →production. This uptake shows a strong seasonal cycle with a maximum during algal →blooms (in spring in temperate areas). Zooplankton grazing on →phytoplankton regenerates substantial amounts of P fixed by algae. Further release occurs during →exudation, →autolysis and bacterial breakdown (→mineralization). Renewed uptake by phytoplankton closes the cycle. Mineralization occurs in the water column as well as in the sediment. The presence of a →thermocline may block the sedimentary P source from the surface waters, but vertical mixing in coastal waters and upwelling will bring this P within reach of the phytoplankton in the →euphotic zone. Locally seabirds bring appreciable amounts of P to land, where their feces (→guano) in breeding colonies may form an economically important source of fertilizer. See also →nutrients.

photic zonation three depth zones can be discerned in the watercolumn according to light penetration. In the upper →euphotic zone (also photic zone or photosynthetic zone) the rate of →photosynthesis by algae is sufficient to compensate for losses due to respiration (both during day and night). The lower boundary is approximately the depth where 1 % of the surface light is found. Below this zone a dimly lighted disphotic zone is found where no effective photosynthesis occurs (algae cannot grow). In the deepest, lightless, →aphotic zone no photosynthesis is possible.

photoautotroph see →metabolic diversity.

photochemical reactions the solar energy flux at the equator surface is equal to about $6 \cdot 10^{21}$ photons cm^{-2} day^{-1} for visible and UV light. Part of it is absorbed to initiate photochemical reactions through radical formations, resulting in intramolecular rearrangement, isomerization, ionization, and intramolecular decomposition. The effect is greatest on →surface films and in the →euphotic zone.

photoheterotroph see →metabolic diversity.

photo-inhibition (syn. light inhibition) the depression in photosynthetic rate of plants at high light intensities (above the light saturation point), resulting from longer exposures and increasing with exposure time. Caused by injury/ destruction by UV of →chlorophyll under high light intensities in the surface waters. See also →light.

photolithotroph see →metabolic diversity.

photo-organotroph see →metabolic diversity.

photoperiod light period, in natural or artificial (experimental) light regime.

photosynthesis light-dependent assimilation of carbon dioxide into cellular organic compounds. In →eukaryotic organisms photosynthesis occurs in cell organelles, i.e., chloroplasts (see →chromatophore). In →prokaryotes (phototrophic bacteria) photosystems are located on membranes in the cytoplasm. With H_2O as electron donor oxygen is produced (oxygenic photosynthesis). Some photosynthetic bacteria can use hydrogen sulfide as an electron donor. The rate of photosynthesis is affected by many environmental factors such as the quality and the intensity of light, temperature and the availability of nutrients. Phytoplankton can adapt its photosynthetic activity to environmental conditions. This adaptation is species-dependent.

photosynthetic bacteria (syn. phototrophic bacteria) see →photosynthesis.

phototaxis movement of an organism in which the directional stimulus is light. Towards the light is called positive phototaxis and away from the light negative phototaxis.

phototroph see →metabolic diversity.

phycobilin plant pigments characteristic for →Cyanobacteria (blue-green algae). Two phycobilins exist (a) phycocyanine, which is blue and gives the blue color; (b) phycoerythrin, which is red and occurs particularly in picocyanobacteria, small picoplanktonic algae abundant in oligotrophic blue ocean waters. See also →pigments, plants.

phytobenthos see →benthos.

phytodetritus see →detritus.

phytolith stony or mineral structure, generally microscopic, secreted by a living plant, e.g., as a defense against →herbivores; often composed of calcium oxalate or →opaline silica.

Phytomastigophora (syn. Phytoflagellata) see →Mastigophora.

phytophagous plant-eating.

phytoplankton the whole group of (usually microscopic) floating plants, mainly →algae. The term refers to a functional group (floating plants) not to a systematic entity, and is used for the →autotrophic part of the →plankton.

picoplankton small (between 0.2 and 2 μm) planktonic organisms, see →plankton.

pigments, animal any material that colors the inside or outside of an animal is called a pigment. However, not all colors in animals are due to pigments; some may be caused by reflections on surfaces which, unlike pigments, absorb no part of the light, and others are due to →bioluminescence. Some pigments originate from pigments of plants used as food: →chlorophyll, →carotene, →xanthophyll, anthoxanthin and their derivates, others come from blood pigments or from their break-down products. Respiratory pigments (→hemoglobin, hemerythrin, erythrocruorin, chlorocruorin, and →hemocyanin) are used for storage and transport of oxygen in the body. Other pigments are used to protect delicate tissues such as nerves from an excess of light which would be fatal; they may function for the detection of light, for absorption of heat rays, or as protective colors (see →coloration in marine animals, camouflage).

pigments, plants substances found in plants which absorb certain wave-lengths of visible light; the absorbed energy is used directly or indirectly for →photosynthesis. Except for →Cyanobacteria these pigments are located in →chloroplasts. There are three groups of pigments: →chlorophylls, →carotenoids and biliproteins. Chlorophyll-a is present in all plants, but each taxonomic group of algae has its additional set of pigments and hence its special color (red, green, brown, blue-green algae). The division of algae into classes is largely based on pigments. In oceanography the concentration of the chlorophylls is often used as a measure of →phytoplankton abundance, while the occurrence of specific pigments can be used as a tool to identify different types of algae. The degradation of algae is accompanied by partial degradation of their chlorophyll, and the degradation products, pheophitins, can be used as an indicator of algal degradation.

pillow lava a general term for →lavas that exhibit a sack-like, bulbous of structure about a meter long; occurring mostly when lava flows underwater.

pinocytosis is the uptake of macromolecules into a cell by a drinking type of action; compare →phagocytosis.

Pisces (fish) aquatic vertebrates (subphylum →Vertebrata, phylum →Chordata) which belong to one of the four classes Agnatha, Placodermi, Chondrichthyes and Osteichthyes. (1) Agnatha (syn. Cyclostomata) are slimy-skinned eel-shaped creatures, which are the most primitive of all living vertebrates. They have no bones, and instead of the backbone there is an elastic →notochord studded with separate pieces of cartilage. All further parts of the skeleton are cartilaginous, including the feeble support of the fins. There are no true jaws, and the gills are found behind one to seven separate pores in the side of the body. To this class belong the lampreys (example: *Petromyzon*) and the hagfish (example: *Myxine*). (2) Placodermi, a fossil group of archaic shark-like and flattened ray-like fish with a stout external skeleton of bony plates. Their remains are found in →Devonian and lower →Carboniferous strata. (3) Chondrichthyes (or cartilaginous fish) are marine fish with a cartilaginous internal skeleton, in particular the fin supports, although the vertebrae may be calcified at the surface. There is no fused skull in the head, and the upper jaws are independent of the brain case. The backbone ends in the upper lobe of the powerful tail fin. There are generally five gill slits on each side behind the head; the skin is covered with sharp, partly embedded tooth-like structures (the dermal denticles). The fertilization is internal, and for this purpose the males have prominent copulatory organs (claspers) on the inner side of the pelvic fins. They are viviparous or have leathery eggs. All existing sharks and rays belong to one Chondrichthyes subclass the Elasmobranchii (syn. Euselachii), the other three containing only extinct forms. Examples are *Squalus* and *Carcharius* (sharks), *Squatina* (monkfish), *Raja, Torpedo* and *Dasyatis* (rays) and *Chimaera* (rattails). (4) Osteichthyes (syn. Teleostei or bony fish) far the largest group containing most of the important foodfish from freshwater

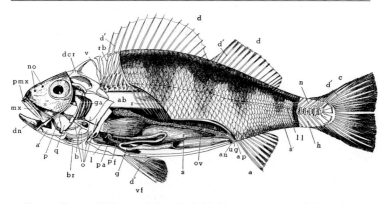

Pisces. (Romer 1962). *a* anal fin; *ab* air bladder; *an* anus; *ap* abdominal pore; *a′* articular; *b* bulbus arteriosus (conus arteriosus); *br* branchiostegalrays; *c* caudal fin; *d* dorsal fins; *d′* dermal rays of fin; *dn* dentary; *dcr* dorsal crest of skull; *g* intestine; *ga* gill arches; *h* hemal spines; *l* liver; *ll* lateral line; *mx* maxilla; *n* neural spines; *no* nasal openings; *o* opercular bones; *ov* ovary; *p* pterogoid; *pa* pyloric appendices; *pf* pectoral fin; *pmx* premaxilla; *q* quadrate; *r* ribs; *rb* basal fins to supports; *s* stomach; *s′* scales; *ug* urogenital opening; *v* evrtebral centra; *vf* pelvic (ventral fin)

and the sea. They are distinguished from the Chondrichthyes by their ossified or bony skeleton. Particularly the skeleton of the head contains many bony elements fused into a skull with well-developed tooth-bearing jaws. There is a single gill opening on both sides, covered by a complex of bones forming the operculum or gill cover. The skin is covered by partly embedded bony plates, the scales, which are thought to have formed during →evolution when fishes spread out into freshwater. A →gas bladder is present which besides providing for the →buoyancy of the fish, is sometimes used as a sort of respiratory organ, or a resonance box for the auditory organs. Bony fish have external fertilization and are, with a few exceptions, oviparous. This class is divided into three subclasses: the Sarcopterygii with the lungfish and the lobe fins, which represent the famous coelacanth, the Brachiopterygii with the bichirs (*Polypterus*) and the Actinopterygii or higher bony fish. This latter subclass with over 20 000 known species is the largest and most diverse group of vertebrate animals. They are a very important source of protein for mankind and millions of tons are harvested each year, providing employment for millions. The most important are the herring (*Clupeia*), salmon (*Salmo*), cod (*Gadus*), perch such as scad (*Trachurus*) and mackerel (*Scomber*), and finally the flatfish (like *Pleuronectes, Solea*).

piscivore (or ichtyophage), animal that feeds mainly on fish. Many seabirds, marine mammals, and most larger fish are piscivorous.

piston corer see →corer.

pitot tube a device for determining flow velocities by measuring a pressure difference between a first orifice facing the flow and a second orifice, located downstream, parallel to the flow, utilizing →Bernoulli's law. As the pitot tube is a symmetric body, the first orifice, at its tip, is at a stagnation point where the flow velocity is zero. As the second orifice (at its downstream side) is at the same (separation) streamline the static pressure measured there will be less than that at the first orifice, since, by Bernoulli's law, the total pressure – the sum of the static pressure and the dynamic pressure $\frac{1}{2}\varrho v^2$ (where ϱ is the fluid density and v the flow velocity) – will be constant. The measured pressure difference thus enables one to deduce the flow velocity: $v = (2(P_{head} - P_{side})/\varrho)^{1/2}$.

plane wave a sinusoidal →progressive wave that travels in one direction, while in the perpendicular direction all points have the same phase. Plane waves are thus parallel, long-crested waves.

planetary vorticity see →vorticity.

planetary wave see →Rossby wave.

planktivores animals feeding on →plankton.

plankton floating organisms of many different phyla, living in the →pelagic of the sea or in freshwater, to a large extent subjected to water movements. (Functional) classification is based on →trophic level, size and distribution. →Autotrophs (primary producers) constitute →phytoplankton, →heterotrophs (consumers) bacterioplankton and →zooplankton. A differentiation in size classes is related to retention by different →mesh sizes of plankton nets and filters used, which, however, have no standard dimensions: →picoplankton <2 µm; →nanoplankton 2–20 µm; →microplankton 20–200 µm; →mesoplankton 0.2–2 mm; →macroplankton >2 mm. In marine plankton the more coastal organisms are termed →neritic, in contrast to oceanic plankton. Related to depth of occurrence is a differentiation in epipelagic to bathypelagic plankton. →Holoplankton has a completely pelagic life cycle, while →meroplankton consists of temporarily pelagic stages of otherwise sessile and →benthic organisms.

plankton torpedo plankton net in a torpedo-shaped housing, to be towed behind a (fast-)moving ship. The basic function of the housing, with a narrow entrance in front of a funnel-shaped widening, is to reduce the inflow, and, by the generated underpressure behind the funnel, to increase the flow through the plankton gauze, which is necessary when fast (up to 6 →knots) hauls are made.

plant production see →production, primary.

plasmid is an extrachromosomal genetic element not essential to growth. Many plasmids carry genes that control the bacterial production of toxins and provide bacterial resistance to antibiotics and other drugs. The genetic information of biodegradation of →xenobiotic compounds such as →herbicides or →pesticides is often located on plasmids which occur in bacteria, so that these strains of bacteria can degrade such new very refractory molecules. Such organisms able to metabolize xenobiotics are of considerable evolutionary interest, since their existence has emerged only over the past 50 years. The construction by bioengineering of plasmids with the genetic information for enzymes to degrade a variety of pesticides or hydrocarbons may help combat environmental pollution with toxic compounds.

plate (tectonic plate) one of the large, rigid or relatively rigid, segments of the →lithosphere, able to move with respect to other plates, and with respect to the underlying mantle. See also →plate boundary, →plate tectonics.

plate boundary the edge of a lithospheric →plate: generally a zone of →earthquake and →tectonic activity, indicating relative motion between plates. There are three types of plate boundaries: (a) convergent, where two plates are moving towards each other. Motion is accommodated by one plate overriding the other. At the surface, converging plate boundaries are manifested as deep oceanic →trenches (resulting from the downward movement of overridden sea floor), or as mountain chains (in zones of continent-continent collision); (b) divergent, where two plates are moving apart, and new oceanic-type →lithosphere is created. These spreading centers show as →mid-ocean ridges or partly as →rifts; (c) →transform, where relative plate motion has no component perpendicular to the boundary and plates are passively sliding against each other. Transform plate boundaries are manifested as →transform faults. See also →plate tectonics.

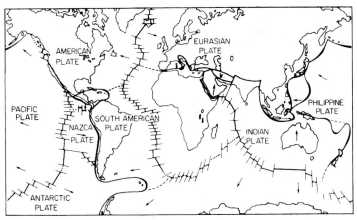

Plate boundary. Location and boundaries of the earth's lithospheric plates. Relative motions shown by *arrows* (assuming a stationary African plate)

plate tectonics a theory was developed in the early 1960s, renewing and founding the hypothesis of →continental displacement. In this concept of plate tectonics, the earth's →lithosphere is divided into a small number of mobile rigid →plates which interact with one another at their →plate boundaries. Along spreading centers, such as →mid-ocean ridges, the plates move apart and new oceanic floor is created by →volcanism. This process is known as sea-floor spreading. Accompanied by minor shallow →earthquake activity, the hot oceanic lithosphere drifts away: aging and cooling with increasing distance from the ridge. Elsewhere, plates move towards each other, one plate overriding another. At so-called subduction zones, oceanic plates sink into the mantle. Subduction is intensified because cooling causes the ocean floor to become denser than the underlying hot mantle. The descending slab is attended with earthquakes systematically distributed down to 700 km below the earth's surface. The downward movement is expressed as a deep oceanic →trench parallel to the plate contact, whereas reabsorption of lithosphere induces →volcanism on the overriding plate (see also →island arc). Continents ride passively along on plates. Where continents meet, subduction is converted into continent-continent collision, crushing the lithosphere into mountain chains. Continents can be split apart and rejoined in this way, but cannot be consumed in subduction zones as in the case of the ocean floor. At transform plate boundaries, lithosphere is neither created nor destroyed; instead plates slide against each other along →transform faults (e.g., the San Andreas fault), giving rise to relatively shallow but sometimes severe earthquakes. See also →Gondwana, →Iapetus Ocean, →Laurasia, →Pangea, →Panthalassa, →Rheic Ocean, and →Tethys.

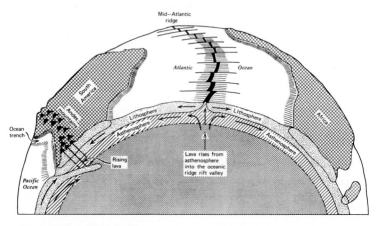

Plate tectonics. (Wyllie 1976)

Platyhelminthes phylum of the two-sided symmetrical flatworms containing the parasitic classes Monogenea and Trematoda (mostly endoparasites of fish), the →Cestoda or tape worms, and the class of the free-living →Turbellaria, which are mainly marine. The body is primitive, unsegmented, and flattened, elongated or oval; the head sometimes has tentacles. The marine forms live mostly in the →interstitial spaces in muddy bottoms, and belong to the →meiofauna.

Pleistocene see →geological time scale.

Pliocene see →geological time scale.

plough mark furrow on the floor of a sea or lake caused by churning and scouring by →icebergs.

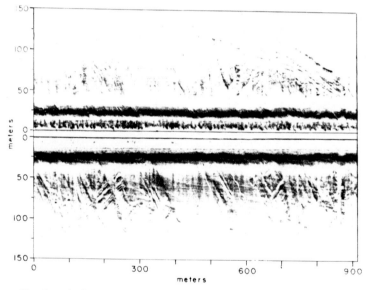

Plough mark. (Reimnitz et al. 1973)

POC see →particulate organic carbon.

pockmark shallow seabed depression, typically several tens of meters across and a few meters deep. Pockmarks are generally formed in soft, fine-grained seabed sediments by the escape of fluids or gases into the water column. (Figure see p. 208)

poikilotherm (cold-blooded animal) organism that is not able to stabilize its body temperature. In these organisms the heat produced by →metabolic

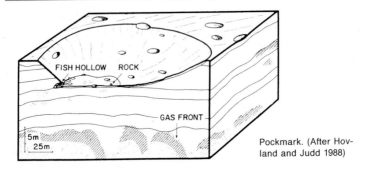

FISH HOLLOW ROCK

GAS FRONT

5m
25m

Pockmark. (After Hovland and Judd 1988)

activity is lost to the environment. As the body temperature closely follows the changes in the environmental temperature, the activity of these animals is also highly variable. At low temperature the animals may become very sluggish. All living organisms, with the exception of most mammals and birds, are poikilothermic. On land some poikilothermic species attain a certain degree of homeothermy (warm-blooded) by habitat selection. On the other hand some homeothermic animals may (temporarily) lose their homeothermic capabilities during inactive periods (e.g., during hibernation).

Poincaré wave see →Sverdrup wave.

poisons substances of natural or anthropogenic origin that have an inherent tendency to destroy life or impair health. Marine animals can be venomous and poisonous, and marine plants may be noxious either to protect themselves against other organisms or as help in capturing prey. The toxic dinoflagellate *Gonyaulax catenella,* when consumed as food by shellfish and then eaten by humans, may cause outbreaks of serious →paralytic shellfish poisonings. Mussel poisonings may cause death to humans from respiratory paralysis. The sting of jellyfish and other coelenterates is caused by the minute stinging capsules or →nematocysts which inject →venom into the tissues of prey or victim. Snails of the genus *Conus* produce a virulent venom that may be lethal to humans upon injection by the animal. It is primarily a mechanism used in feeding but may be used as defence. Some fish are poisonous to eat (poisonous fish); others, such as the stingray, produce venom (venomous fish). See also →venomous marine organisms and →toxicity.

polar wandering displacement of the position of the geomagnetic pole in relation to its present position. It is possible to measure the direction of magnetization of suitable rocks and to infer the position of the geomagnetic pole at the time of their formation (see also →magnetic (earth) field). If it is assumed that the earth's magnetic field has always been dipolar, more or less fixed in direction relative to the geographic poles, the implication of

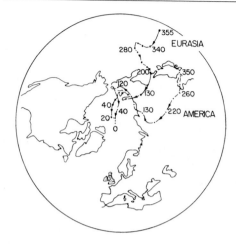

Polar wandering. (After Holmes 1965). Position of the pole for the American and Eurasian continents during the last 355 million years (*numbers* indicate million years)

polar wandering is that during geological time continents systematically occupied other positions. See also →continental displacement and →plate tectonics.

pollen microgametophytes of seed plants with a very resistent organic outer wall (exine); used for →stratigraphic and →paleoecologic purposes; widely distributed by air and water.

pollutant any chemical substance from outside the system, whether of natural or anthropogenic origin, with an impact on biological processes in the system. See also →pollution and →contaminant.

pollution (marine) is the introduction by man, directly or indirectly, of substances or energy into the marine environment (including estuaries), with such deleterious effects as: harm to living resources, hazards to human health, hindrance to marine activities, impairing the quality of seawater, and reduction of amenities (as defined under the Intergovernmental Oceanographic Commission, IOC). Pollution effects are exerted at any integration or organization level in the natural environment: on molecular and cellular levels, on physiological processes in organs and organisms, on behavior of individuals, populations or separate parts of them, on habitats and ecosystems. Pollution legislation means making or enacting laws concerning pollution; or a law or a body of laws enacted, with respect to pollution. Thermal pollution effects are due to unnatural, man-made temperature (thermal) changes in a system; radioactive pollution is the presence of either natural or anthropogenic radioactive substances with toxic or nuisance effects. When there is no noticeable effect of these substances, the term contamination is applicable. Radioactive pollution, or contamination, of the marine environment is of four types: →fallout (radioactive) from bomb-test and nuclear experiments, dumping of solid low-level wastes, liquid effluents from nuclear reprocessing plants and cooling water of nuclear power stations. Fallout precipitates have globally been the major contamination, but are gradually decreasing due to decay since 1963, when experiments of this kind were practically stopped. Dumping in the sea of solid wastes has also been halted due to national and international political pressure. Liquid effluents from reprocessing plants, like those in Sellafield (UK) and La Hague (F), consistently contain radionuclides, which are released under strict national regulations. In principle, nuclear power plants may only release cooling water which contains for instance small amounts of neutron-induced radionuclides not exceeding permissible levels.

poly- prefix meaning many.

Polychaeta (polychaetes) class of the phylum →Annelida or segmented worms, perfectly →metameric, with cylindrical body segments, a pair of fleshy appendages called parapodia, with many →chaetae. The polychaetes are a vast class (more than 5000 species) containing most of the marine annelids. They can be free-moving or errant, or sedentary living in tubes. Some live in the →pelagic, others in gravel or sand, or in burrows in muddy sediments (e.g., *Arenicola marina*). Polychaetes show a variety of feeding types and accordingly specialized anatomy, such as collecting particles from the water over the burrow or tube with a crown of fine tentacles, crawling over the bottom and collecting food (algae, animals, carrion, or combinations), or they live in the bottom as →deposit feeders. (Figure see p. 211)

polychlors the term polychlors includes organochlor compounds, among which those of DDT (and metabolites such as DDD, DDMU, and DDE), dieldrin, endrin, aldrin (γ-HCH). PCBs (polychlorobiphenyls) are the most common as →contaminants in the sea. Since the ban on DDT, dieldrin and endrin as pesticides in most of the developed countries, concentrations in all environmental compartments have decreased. In tropical areas, however, where these compounds are still applied in antimalaria programmes (DDT) and to fight swarms of locusts (dieldrin), environmental contamination still occurs. PCBs are a group of chlorinated aromatic compounds, produced synthetically by chlorinating the biphenyl skeleton. Theoretically, 209 different chlorobiphenyls can exist based on the possible distribution of Cl atoms in the two rings of the biphenyls. The amount of chlorine in PCBs ranges from ca. 18 to 71 %, and each technical mixture contains at least 37 biphenyls with different degrees of chlorination (congeners), detectable with capillary gas chromatography with electron-capture detection. Their application was primarily in the field of noncombustible lubricators and isolation oils in electrical transformers, but also as additions to paint and no-carbon-required paper. Since 1945, total world production has been

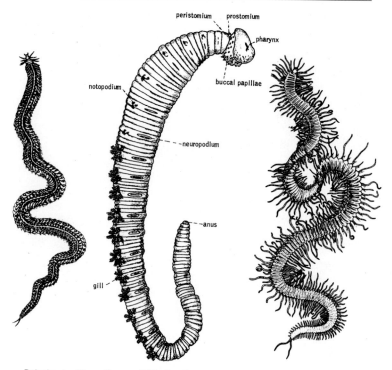

Polychaeta. (From Barnes 1987 after Brown 1950)

about 1.2 10^6 ton y^{-1}. Since 1980, PCBs have gradually been banned, but because these compounds are very persistent, their effect on the environment will persist for many decades. In 1988, 31% of the total production was dispersed in the environment, while 65% is still in use, or otherwise potentially able to reach the environment. Polychlors are hydrophobic compounds and therefore accumulate mainly in lipids of aquatic organisms, either directly from water or through food. In aquatic organisms equilibria are attained between concentrations in these lipids and water with →distribution coefficients of about 10^4 to 10^6. Accumulation along the food chain is not evident. This is different for sea mammals and sea birds, which lack a system for exchange with seawater, and therefore may accumulate much higher concentrations. Metabolic conversion of polychlors is possible for DDTs and some low-chlorinated PCBs.

Polynya open area between the polar sea ice fields (see →ice).

polyp general name for individual soft-bodied animals of polypoid shape, i.e., of hollow cylindrical shape, with central mouth opening, surrounded by tentacles, standing on a foot. Polypoid individuals occur in the →Coelenterata, as sessile vegetative life form of the →Cnidaria, in →corals, and in →Bryozoa.

polyphagous thriving on many different food types.

Polyplacophora chitons, class of the phylum →Mollusca. Mollusks with a flattened, oval body and a strong foot, to adhere strongly to hard substrate, and a heavy mantle with a dorsal shell, typically divided into eight overlapping shell plates. There are some 500 species, living in rocky intertidal areas, where they overtide, clinging to rocks and stones, and start crawling and feeding when submerged. They feed by scraping algae, etc. from the substrate with the →radula.

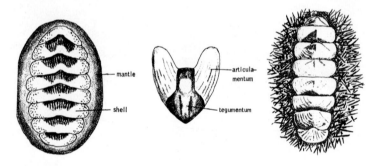

Polyplacophora. (Barnes 1987)

polysaccharides belong to the group of →carbohydrates. Mono- and disaccharides are sugars, while polysaccharides are products such as starch, cellulose, and glycogen. In the marine environment these substances are detectable at low concentrations mainly in surface layers with a high primary productivity.

Polythalamia →Foraminifera with a test/shell of several chambers. Example: *Globigerina*.

Polyzoa see →Bryozoa.

population all individuals of a species living in a certain area. Populations may be either closed (i.e., there is negligible interchange with other populations of the same species) or open. The size of a population fluctuates, increasing by reproduction (→natality, →recruitment) and →immigration, and declining by death (→mortality) and →emigration. In →population dynamics, population size is usually expressed as population density, i.e., the number per unit of surface (n m^{-2}) or volume (n m^{-3}).

population dynamics the study of animal and plant populations of a single species, the patterns of change in space and time and the factors affecting them. The dynamics in time are determined by →birth and →mortality rates, the dynamics in space are determined by migratory processes. →Biotic as well as →abiotic factors continuously affect these processes, thus causing the population to fluctuate. Some factors are important in controlling these processes. Other factors act primarily as negative feedback, resulting in a reduction of the scale of population fluctuations and maintaining the population at or about an equilibrium size, the carrying capacity. Populations very often have an age structure: besides the adult stage of an organism, one or more juvenile stages can be distinguished which differ in size, age and behavior. In such populations only the adults have reproductive capacities. An important property of such a population is its generation time: the time between the appearance (birth) of the most juvenile stage and the moment that this cohort of organisms has reached the adult stage and has produced offspring.

population parameters (1) (math.) describe the properties of a mathematically defined population and include population mean (μ), variance (s^2) and standard deviation (s). Variance and standard deviation are a measure of variability in the population. Population parameters can be estimated by corresponding statistics (see →statistics); (2) (biol.) the specific values which characterize a population in time and space. They concern density or →abundance, →age distribution, birth rate or →natality, death rate or →mortality, and →carrying capacity.

Porifera phylum sponges. These are the most primitive of the multicellular animals, lacking any organs or tissues, but in principle consisting of a structure of cells with pores, arranged around a spongocoel or lumen, and supported by mesenchyme (primitive connective tissue) containing skeletal spines (calcareous or silicious spicules or spongine fibers). Through the pores, water with oxygen and food is pumped by →flagellar beating from outside into the spongocoel, from where it leaves the sponge. There is no organization in the activities in sponges; any part of the body can regenerate. They are →hermaphrodites; fertilized eggs produce flagellated larvae which become attached to form new sponges. The forms can be simple to very complicated with networks of canals and chambers. They live sessile in marine and freshwater. Porifera are divided into the classes Hexactinellida or Hyalospongiae (glass sponges, with six-pointed siliceous skeletal spicules), the →Calcarea or Calcispongiae (calcium carbonate spicules), the →Demospongiae (siliceous spicules, spongine fibers or both)

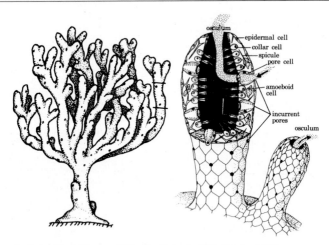

Porifera. (From Barnes 1987 after Reiswig 1975; Keeton 1972)

and the Sclerospongiae (with an external covering of calcium carbonate). The common bath sponges with skeletons of exclusively spongin fibers belong to the family Spongiidae (class Demospongiae).

Portuguese man-of-war see →Siphonophora.

potential flow (syn. irrotational flow) flow free of →vorticity that can therefore be described as the gradient of a scalar function, the velocity potential.

potential temperature according to the laws of →thermodynamics, water that moves →adiabatically upward or downward, decreases or increases in temperature respectively, because of the change in pressure and the →compressibility of seawater. Therefore temperatures at different depths are not directly comparable. The potential temperature is defined as the temperature that a parcel of water would have if it were moved adiabatically upward to the surface. This value, independent of depth, makes comparison possible. The potential temperature (symbol θ) can be calculated from the →in situ temperature of deep water, salinity, and pressure (say: depth). The potential temperature is slightly lower than the temperature in situ.

Prasinophyceae class of the division →Chlorophyta, planktonic algae, usually with four →flagella, sometimes two or one. Typical are the small organic scales, which were used to separate them from the →Chlorophyceae. However, comparable scales have now been found in some Chlorophyceae, which makes separation of the Prasinophyceae doubtful.

Precambrian see →geological time scale.

precipitation the separation of a substance in solid form from a solution by means of a reagent. In meteorology, the falling products of condensation in the atmosphere, such as rain, hail, or snow. Precipitation at sea is difficult to measure because the rain gauges used on land, if used at sea, are affected by the wind disturbances around the platform and may also sample sea spray. The precipitation values at sea are therefore mainly based upon indirect estimates.

predation-prey relationship there are several predator-prey relationships in a →community. The dynamics of both populations react on each other. The number of prey animals eaten per predator per time unit increases with prey density (functional response), as does the growth rate of the predator population (numerical response). The impact of predation on prey →mortality depends on both responses. Hence predation pressure and rate of increase of the predator population are correlated with prey density, which can lead to density oscillations in both populations.

predatory feeding see →raptorial feeding.

preservation (1) marine foods can be preserved deep frozen, dried, salted, smoked, cooked, tinned, etc; (2) for scientific purposes (marine) organisms can be preserved in chemicals like alcohol and formalin or deep frozen; (3) populations can be preserved by limitation of their exploitation. Local, national, and international regulations are necessary to preserve →habitats and (migratory) animals and plants.

pressure force per unit of surface, expressed in pascal (Pa) $= 1$ newton m^{-2}, but in oceanography often in atmosphere (atm) or (deci)bar (1 bar $= 0.987$ atm $= 10^5$ Pa). The pressure in the sea is related to depth by the →hydrostatic equation. Usually the atmospheric pressure is assumed to be constant in the estimation of the subsurface pressure. Pressures in the oceans are often given in decibars as 1 decibar equals roughly 1 m increase in depth. At the high pressures (> 1000 decibars) in the deep ocean some chemical effects occur because equilibrium reactions will seek the smallest possible volume, thus counteracting compression. Those known are limited to the bicarbonate system and saturation of →calcium carbonate as related to →compensation depth. Effects of pressure on deep-sea animals must be of great physiological importance, but observations are few. Shallow-water organisms usually do not survive several hours of pressures encountered in the →abyssal zone (> 200 atm. at > 2000 m depth).

pressure gauge instrument used to determine the depth of an instrument below the sea surface, or, alternatively, when mounted at the sea bottom, to determine sea-level variations. An early method for measuring the depth of an observation by means of a pressure sensor is the combination of two →reversing thermometers, one of which is not protected against pressure and therefore gives a higher temperature reading. This deviation can be used to calculate the pressure (depth).

Priapulida phylum of only 13 known species of the →Aschelminthes. Cucumber-shaped or worm-like animals living buried in the sea bottom. The animals consist of a spiny →proboscis (one-third of the body) for catching soft-bodied prey, and an anterior part or trunk. They live in soft bottoms in colder seas of the northern hemisphere.

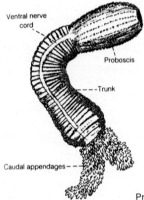

Ventral nerve cord

Proboscis

Trunk

Caudal appendages

Priapulida. (From Barnes 1987 after Hyman 1951)

primary production see →production, primary.

primary wave see →P-wave.

prions (1) small proteins apparently self-replicating but completely devoid of nucleic acid. See also →virus; (2) group of, particularly Antartic, sea birds.

proboscis tubular process of the head, used in feeding, burrowing, or locomotion.

procaryote see →prokaryote.

producers all organisms which are able to synthesize organic material from less complex compounds such as →carbon dioxide. All green plants belong to this →trophic level and produce organic material by the process of →photosynthesis. This production process is also called →primary production. Secondary producers are not producers in this sense, but convert vegetable organic material obtained from their food into their own body material. Therefore, they should properly be called →consumers.

Tertiary producers are consumers of animal material. They are at the top of the →food chain.

production, efficiency ratio between secondary and primary production or more general: between production at trophic level $n+1$ and level n (see →trophic level and →food chain). This ratio is always smaller than 1.0, because not all level-n production is consumed by level $n+1$ and not all consumed material is used for growth (parts of it are respired and excreted).

production, primary the rate at which energy is stored by photosynthetic or chemosynthetic action of producer organisms, in the formation of organic substances (which can be used as food). →Photosynthesis is a fundamental process in primary production; primary producers are chiefly green plants; in the sea they are microscopic (→phytoplankton). In the sea, nutrients, light and sometimes CO_2 may be →limiting factors in primary production. Primary production is usually expressed as g C per unit of area and time. Gross primary production is the total rate including organic matter used by the primary producers in respiration during the measuring period. Net primary production is the excess of primary production over respiration of the primary producer during the measuring time, i.e., the rate of storage of organic matter in the primary producer.

production, secondary the rate of production of →herbivorous animals by conversion of their vegetable food into animal tissue, see →producers. It is generally expressed in mass units per unit of area per unit of time, e.g., $g\ m^{-2}\ y^{-1}$. It is usually estimated by summing mean mass increments of individuals occurring in known numbers per unit area over a period of time.

production, tertiary the rate of production of →carnivorous animals by conversion of their animal food into their own tissues. See also →producers.

production-respiration ratio ratio between organic material synthesized by an organism and the amount respired. See further →producers and →production efficiency.

productivity relative rate of →production.

proglottids segments of the tapeworm; when mature each proglottid has at least one set of reproductive organs.

progressive vector diagram representation of the variation in current velocity with time. The velocity at each moment is drawn as a vector on top of the vector of the earlier time step. The succession of vectors forms a diagram that represents the mean (→Eulerian) movement for the time series

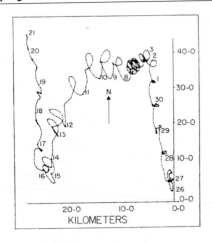

Progressive vector diagram.
(After Le Blond and Mysak
1978)

and its variations. The picture looks like a trajectory (see →Lagrangian representation), but is a representation of the velocity at one point.

progressive wave wave moving relative to the fluid and transporting energy in that direction. Progressive waves (or travelling waves) are usually considered →plane waves of a sinusoidal shape. They can be expressed mathematically by a wave-number vector, pointing in the direction of propagation and with a length equal to the wave number (see →wave characteristics). The counterpart of progressive waves is →standing waves, in which there is no phase propagation.

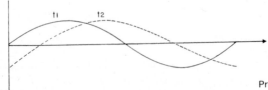

Progressive wave

prokaryote a cell or an organism lacking a true nucleus, usually having its →DNA in a single molecule and not organized in chromosomes. →Bacteria and →Cyanobacteria belong to the prokaryotes.

Prosobranchia subclass of the class →Gasteropoda (phylum →Mollusca), which undergo torsion during larval development; nearly always with a shell and an →operculum. They have a visceral loop twisted in a figure of eight, the mantle cavity opens anteriorly, the →ctenidia are in front of the heart. Sexes are separate. The majority of the Gastropoda belong to this subclass.

protein(s) extremely complex nitrogen-containing organic compounds of numerous →amino acids found in all animal and vegetable cells, where they constitute a major part of the living protoplasm. They differ from →carbohydrates and fats not only in their function in the organism, but also in elementary composition. In addition to carbon (C), hydrogen (H), and oxygen (O), proteins invariably contain nitrogen (N) and generally also sulfur (S). The percentage composition of the large number of proteins falls between rather narrow limits: C = 50 to 55 %, H = 6 to 7.3 %, O = 19 to 24 %, N = 13 to 19 %, and S = 0 to 4 %. Proteins may also contain phosphorus, iron (see →hemoglobin), copper (see →hemocyanin), iodine, manganese, zinc, and other elements. None of these elements, except iodine, has so far been found as a constituent of the →amino acids, the fundamental units of which proteins are built up by the organism. A protein molecule may consist of hundreds or thousands of amino-acid molecules, which are usually of about 20 different kinds. The amino acids are linked together by →peptide links. The possible different arrangements of the amino acids are evidently practically infinite, and the diversity is fully exploited by living beings, every species, possibly every individual, having kinds of protein molecules peculiar to itself. A protein molecule is very large (most have a molecular weight from about 20000 up to several million), and dissolved proteins therefore form colloidal solutions. Proteins are not soluble in fat solvents. Many are soluble in water or dilute salt solutions (e.g., globulins); others, with elongated (fibrous) molecules, are insoluble in these solvents (e.g., scleroproteins, myosin). Proteins are synthesized by all living beings, but only →autotrophic organisms make them from inorganic substances. Proteins are frequently combined with other substances, especially nucleic acid (nucleoproteins), carbohydrates (muco- or glycoproteins), and fats (lipoproteins).

Protista (protists) see →Protozoa.

Proto-Atlantic Ocean see →Iapetus Ocean.

Protobranchia the most primitive order of the →Bivalvia in which the →ctenidia are unspecialized and retain their primitive function of respiration. The →labial palps are enlarged as a feeding mechanism. The foot is anteriorly bilobed. There are two adductor muscles. Example: *Nucula*.

protoplast see →spheroplast.

Proto-Tethys see →Rheic Ocean.

Protozoa (protozoans) general term for a wide variety of microscopic unicellular organisms (ca. 50000 species), most of which are motile and →heterotrophic. The Protozoa were considered to form one phylum within the animal kingdom, though some are →autotrophic and contain →chlorophyll and other →pigments), which made it very difficult to decide whether they are animals or plants (algae). At present, the different groups of

Protozoa are treated as separate phyla, which belong together with most algal phyla to the kingdom →Protista (in which the difference between plants and animals is irrelevant, the main characteristic being the unicellular level of organization). Protozoa live in the sea, in freshwater, or as parasites inside other organisms. They usually multiply by fission (division of the cell into two or more new cells), though sexual reproduction exists as well. The different phyla are distinguished according to the types of locomotory organelles (specialized parts of the cell functioning as an organ in multicellular organisms). (1) Sarcomastigophora, with locomotion by →flagella or pseudopodia, divided into the subphylum →Mastigophora or Flagellata with the groups →Euglenida, →Volvocida (both considered by botanists as belonging to the →Chlorophyta or green algae), the Dinoflagellata, (see →Dynophyceae), and some other groups, and the subphylum →Sarcodina (locomotion by pseudopodia), divided into →Amoebida, →Foraminifera, →Heliozoa, →Rhizomastigina and →Radiolaria, (2) and (3) Apicomplexa and Microspora, together known as the Sporozoa, parasitic protozoans living within or between the cells of the animal host, and (4) →Ciliophora, ciliates, the largest phylum of the Protozoa (more than 7000 species), which move and take their food with →cilia.

proximal situated near point of attachment, opposite of →distal.

Prymnesiophyceae see →Haptophyceae.

pseudofeces see →ctenidium.

pseudoneuston see →neuston.

PSP see →paralytic shellfish poisoning.

psychrophile (syn. cryophile) able to grow at low temperatures.

Pterobranchia see →Hemichordata.

Pteropoda marine →Gastropoda highly adapted to a →pelagic life-style ("sea butterflies"), naked or with shells composed of →aragonite. The latter is biostratigraphically useful in local correlations of Quaternary deep-sea sediments. See also →microfossil, biostratigraphy and →deep-sea sediments. (Figure see p. 221)

Pulmonata see →Gastropoda.

pumice excessively cellular →lava, generally of the composition of rhyolite (fine grained granite). The cellular structure results from a disruption of the →magma by expanding gas. Some pieces of pumice have enough →buoyancy to float on water.

pure culture (syn. axenic culture) culture of organisms in the absence of any other organisms; it is an uncontaminated culture.

purse seines see →nets.

Pteropoda. (Hardy 1958)

putrefaction protein degradation under →anaerobic conditions. Usually this proces does not lead to an immediate liberation of all the amino nitrogen as →ammonia. Some of the →amino acids are converted to evil-smelling compounds, notably the amines. In the presence of air, the amines are oxidized with the liberation of ammonia.

P-wave (syn. primary wave, compressional wave) a type of body wave (see →seismic wave) that involves particle motion in the direction of propagation (alternating compression and expansion, longitudinal wave). It is the fastest of the seismic waves, traveling 5 to 7.2 km s^{-1} in the crust and 7.8 to 8.5 km s^{-1} in the upper mantle. The P stands for primary; it is so named because P-waves are the first arriving at the observation point.

pycnocline layer in which the →density of the water rapidly increases with depth, because of the presence of a →halocline or a →thermocline or both. See →stratification.

pyranometer (syn. solarimeter) instrument quantifying the amount of solar radiation that reaches the surface of the earth.

pyrite FeS_2, is found in large crystals which give off sparks when struck against metal. This property gave the mineral its name. It is formed diagenetically (see →diagenesis) under anoxic conditions from Fe (II) and sulfur which are both produced during oxidation of organic matter when oxygen and nitrate are exhausted. Especially in coastal regions with enhanced productivity there is enough sulfate to produce pyrite. It is not stabel in the presence of oxygen; the mineral is slowly oxidized to Fe (III) and sulfuric acid. This process leads to very acidic soils after land reclamation in coastal zones (catclay soil).

Q

Q10 the temperature coefficient giving the increase of a reaction rate at a temperature increase of 10 °C. The temperature coefficient of enzyme reactions is 1.4 to 2.

Quaternary see →geological time scale.

R

rad former unit of radiation dose, see →gray.

radiance the amount of light per unit area per unit solid angle from a certain direction. Typically, measurement units are watts per square centimeter per steradian.

radiation is used in different ways in oceanography. In the more common use it refers to electromagnetic radiation (light, infrared radiation, see →optics), or to radioactivity. It may also refer to other forms of energy transport, in particular by →gravity waves.

radiation, background radiation caused by natural sources: cosmic radiation and terrestrial radiation. Cosmic radiation constantly impinges upon the earth; it originates within our own galaxy and at times of intense solar activity its contribution from the sun increases considerably. Terrestrial radiation stems from such naturally occurring isotopes such as ^{40}K, ^{87}Rb, and nuclides from the uranium and thorium series. Background radiation causes "noise" in measuring, e.g., radioactive →tracer activities and is therefore usually shielded off by lead.

radiation stress see →momentum density.

radio tracking see →telemetry.

radioactive fallout see →fallout.

radioactive wastes see →ocean dumping.

radioisotopes or radionuclides are isotopes emitting alpha, beta or gamma radiation from their nucleus. The →half-life ($\tau_{1/2}$) is a constant for a particular nuclide, not dependent on any physico-chemical process. This property makes radionuclides ideal for the determination of time scales, ages, and rates by radiometric dating. From the measured and initial concentrations, the decay after the incorporation of the isotope is determined, and the time is calculated that has elapsed since incorporation took place. The most useful range for any nuclide is from 0.2 to 5 times $\tau_{1/2}$. According to their $\tau_{1/2}$ and sources, radioisotopes can be divided into three different categories: (a) isotopes with a very long $\tau_{1/2}$, of the order of the age of the earth: ^{232}Th, ^{238}U, ^{235}U, ^{40}K, ^{87}Rb. To this category belong also series of daughters, granddaughters, etc. (e.g., all isotopes of Th, Pa, Ra, Rn and ^{210}Pb), which are also radioactive. All these nuclides occur naturally; (b) isotopes formed by interaction of cosmic rays with atmospheric elements. The inventory of these nuclides, ^3H (12.26 y), ^{10}Be (1.5 10^6 y), ^{14}C

(5730 y), ^{32}Si (130–180 y), is a function of cosmic-ray intensity and $\tau_{1/2}$. The testing of thermonuclear devices around 1960 in the atmosphere has greatly increased the inventory of ^{3}H and to a lesser extent that of ^{14}C. ^{3}H is extensively used as a transient tracer in circulation studies; (c) isotopes resulting from the use of nuclear fission; all Pu and Am isotopes and, e.g., ^{90}Sr (28 y) and ^{137}Cs (30 y) fall in this category. The use of ^{137}Cs and ^{134}Cs in water circulation studies around the British Isles is an inadvertent application of the waste discharge of the Windscale/Sellafield processing plants. Of nearly all elements radioactive isotopes are known which are prepared on a laboratory scale and used in the determination of rates. See →absolute age, isotope, →carbon radioactive and →cosmogenic nuclides.

radiometric dating see →carbon, radioactive, →radioisotopes and absolute age.

Radiolaria marine free-living protozoans (Protista) with pseudopodia, classified by some with →Foraminifera and →Amoeba under the Rhizopoda, by others with acantharians and →Heliozoa under the Actinopoda. The living matter or cytoplasm is divided by a mucoid or chitinous layer (the central capsule, only present in Radiolaria) into extracapsular and intracapsular cytoplasm, the latter containing the nucleus. Distinction of the species is largely based on skeletal characteristics; in the spumellarians the skeleton is symmetrical, in the nassellarians helmet-shaped. A few primitive genera lack skeletons. Most of the skeletons are composed of opaline silica and therefore have a fossilization potential. The related acantharians have skeletons composed of strontium sulfate, never preserved in sediments. The freshwater Heliozoa have a radial symmetry resembling spumellarians, but lack a preservable skeleton. Particularly deep-sea sediments may be rich in radiolarian skeletons. Radiolaria have existed since the →Cambrian and probably even earlier. Some seem to extend throughout the whole post-

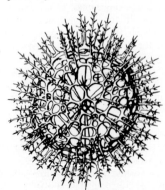

Radiolaria. (Barnes 1987)

Precambrian; only few genera have a limited vertical range. This diminishes their use in →stratigraphy, but locally they have proved useful in correlation of deposits or deep-sea cores, particularly when other fossils are lacking.

radionuclides, natural two groups of natural →radioisotopes exist, viz. →cosmogenic nuclides and primordial nuclides. →Cosmogenic nuclides are constantly produced in the stratosphere, where primordial radioisotopes are the remainder of long-lived nuclides from the formation of the earth, such as potassium (^{40}K) and the isotopes of the ^{238}U, ^{232}Th and families. Daughters of the uranium and thorium families are useful tools for oceanographic studies such as age determinations of sediment layers (^{210}Pb, $^{230}Th/^{232}Th$), growth of corals ($^{234}U/^{238}U$) and transport of terrestrial material to the sea ($^{234}Th/^{232}Th$).

radio-tags see →tagging.

radula "tongue" or horny strip with, up to, thousands of teeth on the surface in the mouth of mollusks. In gastropods especially, a highly developed feeding organ used as a grater, rasp, cutter, grasper, etc.

range see →biohorizon.

raptorial feeding (syn. predatory feeding) feeding like a "raptor" (robber) used, e.g., for birds of prey; in fish and plankton raptorial feeding stands for taking prey one by one, as opposed to →filter feeding.

rare earth elements also lanthanides. Elements lanthanium (^{57}La) through luretium (^{71}Lu). The outer electron shells of the commonly occurring (III) cation exhibit the same electron configuration, hence the very similar physical and chemical properties. The gradual filling with electrons of the 14 positions in the inner 4f shell leads to a gradual change in properties with increasing atomic number. At exactly half filled 4f shell, as for gadolinium (Gd), energetic advantage may yield anomalous properties. Concentrations in seawater ranging from about 10 pmol l^{-1} (^{57}La) down to about 0.1 pmol l^{-1} (^{71}Lu). When normalized to crustal abundance the seawater values increase gradually with increasing atomic number. The isotopic signature of neodymium (Nd) varies between ocean basins. Exist in trivalent (III) cationic oxidation state in nature, except for cerium (Ce) and europium (Eu), which may also occur in tetravalent (IV) and divalent (II) oxidation states.

rarefaction see →diversity.

ray path is a line following the propagation direction of the waves, and is everywhere perpendicular to the wave crests. Between two adjacent ray paths the available →wave energy is conserved. So, if two ray paths are converging, the energy per crest length (wave action density) must increase, and the reverse occurs for diverging paths. Because of →refraction, ray

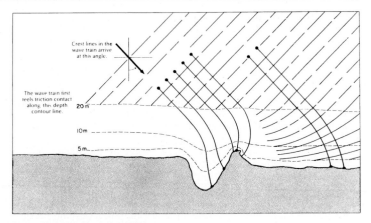

Crest lines in the wave train arrive at this angle.

The wave train first feels friction contact along this depth contour line.

20 m

10m

5 m

Ray path. (After Neshyba 1987)

paths converge to ridges and diverge in gullies; the resulting change of energy influences the sedimentary regime.

Rayleigh number a nondimensional number, Ra, named after the British physicist Lord Rayleigh (1842–1919), that gives the ratio between the →buoyancy force and the combined opposing effects of viscosity and diffusive transport of buoyancy. If the Rayleigh number exceeds a certain critical value, the fluid becomes unstable and convection sets in. Usually the buoyancy will result from heating, and the thermal expansion coefficient is incorporated in Ra, but when buoyancy is connected with salinity differences (as in →estuaries), the volume contraction coefficient for salinity will be incorporated in a Rayleigh number for salinity.

recruitment "upgrowth", term used in population biology for the appearance of the young in a given population per unit of time. The term is also used in a slightly different way, i.e., not referring to the natural population in a given area, but to the catchable population for a particular fishery: recruitment to the catches is the appearance of the year's →0–group in the catches. The size or age at which this occurs depends on →mesh size.

red algae see →Rhodophyceae.

red clay (also brown clay) fine-grained and bright to reddish brown oceanic deposits at depths greater than 5000 m. In these deposits much of the biogenic →carbonates and →opal is dissolved, and the remaining component is clay derived from wind-blown particles, meteoric and volcanic dust, →manganese nodules, ice-rafted debris and diagenetic

→smectite. The red color is mainly due to ferric (Fe^{3+}) iron and manganese compounds. In areas with red clays the deposition rate is less than 1 to 2 mm ky^{-1}. The red clay covers 10^8 km^2, mainly in the Pacific and Indian Oceans.

red tide a red or reddish brown discoloring of surface waters, most frequent in coastal regions, caused by concentrations (→blooms) of plankton, particularly →dinoflagellates. Sometimes →mass mortality occurs due to poisonous substances produced by the algae, or to low oxygen concentrations due to high amounts of oxygen used for decomposition of the algae at the end of the bloom.

Redfield ratio elemental analysis of phytoplankton has revealed that a number of major elements are present in these organisms in a characteristic atomic ratio: $O : C : N : P \sim 276 : 106 : 16 : 1$. Obviously, different mixtures of algae species have a very similar chemical composition of carbohydrates, proteins, pigments, etc. Not only in the plankton itself, but also in the deep water column of the oceans do these nutritional elements, subsequent to phytoplankton degradation, occur in the dissolved state in this ratio. The name of this characteristic ratio is associated with the American oceanographer A.D. Redfield, who contributed significantly to the understanding of the occurrence of a permanent relationship between concentrations of different nutrients.

redox potential the energy required by all forms of life is ultimately derived from the process of electron transfer. Hydrogen atom pairs (equivalent to $2H^+ + 2e^-$) are removed from certain intermediates of carbohydrates, fat, and protein degradation, and fed into the electron transport chain, donating their electrons to carrier components of the chain while the associated H^+ ions enter the aqueous environment. The electron transport chain comprises redox couples whose redox potential displays a small difference from its neighbors; consequently, the transfer of electrons is associated with the release of free energy. The free energy change (G) of the reaction involving electron transfer from a donor couple to an acceptor couple is given by the equation $\Delta G = -nF\Delta E$, where n is the number of electrons transferred, F is the Faraday constant (equal to $96\,486$ coulombs) and ΔE the difference in electrode potential between the two redox couples.

reduced gravity see →buoyancy.

reducing conditions environments where in the absence of oxygen (bio-)chemical processes occur with other electron-acceptors such as nitrate (denitrification), iron oxide and manganese oxide, sulfate (sulfate reduction). Reducing conditions are in general exceptional in the water column (only in stagnant basins like the Black Sea, Cariaco Trench, etc.), but are common in the sediment, where below the restricted penetration depth of oxygen reducing conditions prevail.

reef elevated structures on the seabed built by calcareous organisms. In the past →sponges and →bryozoans were important; the most important modern reef-building organisms are stony corals. Although reef-like structures occur in deep water, e.g., in the northeastern Atlantic, most reefs occur in shallow tropical marine environments and these are characterized by high rates of carbonate accumulation. Coral reefs grow in warm, clean, shallow waters along many tropical coastlines and are built up principally of coral, coralline algae, and debris of →mollusks and →foraminifers. They disappear (a) where water temperatures are below about 14 °C; (b) at depths where the sunlight does not penetrate; (c) where there is freshwater run-off, and (d) where the water is muddy. A common classification lists reefs as fringing reefs (reefs directly lining shores), barrier reefs (developed parallel to the shore but separated from it, e.g., the →Great Barrier Reef in Australia) and atolls (circular reefs surrounding deep lagoons). Fringing reefs occur where a hard, stable substrate is present offshore (e.g., an island or a rocky headland). Barrier reefs grow on the continental shelf at some distance from the coast. Submarine reefs have a reef-front or fore-reef where corals grow luxuriously and fish and many other organisms inhabit crevices and caves. The reef is usually topped by a reef crest which is exposed during low tide and usually contains a large (calcareous) algae population. The channels between reefs (interreef channels, or off-reef floor) have a substrate of unconsolidated, usually rippled sediment. The reef crest can merge into an intertidal reef flat with often deep pools, flourishing coral and coarse sediment. Reef flats often (partly) enclose a shallow lagoon, covered with calcareous mud and patchy reefs. Beaches (cays) are formed from gravel- or sand-size coral fragments; beach rock is often present. On intertidal muds →mangrove may develop. Coral islands (raised reefs) consist of pre-Recent reef rock. Noncoral reefs. Several other groups of organisms build reef-like structures in intertidal and shallow sub-tidal environments: vermitid →gastropods in combination with coralline algae, oysters, polychaete worms (serpulids) and Sabellaria. They form carbonate structures except the Sabellaria, which construct tubes of cemented sand grains and shell debris. Oyster reefs are formed at →salinities of 15 to 25 S in tropical and temperate lagoons, bays and estuaries. The others are formed at approximately seawater salinities; serpulid reefs may occur at salinities between 4 and 55 S.

reflection shooting (seismic) (geol.) a →seismic survey technique based on measurement of travel times of →seismic waves that are reflected at near-vertical incidence from boundaries separating media of different elastic wave velocities and densities. At sea, a cross sectional display of reflection time versus distance along a traverse (or profile) is obtained by towing an acoustic source behind a ship (→seismic profiling). The source emits pulses

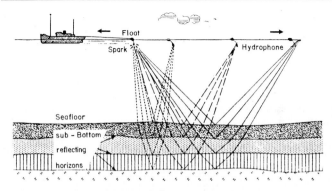

Reflection shooting (seismic). (After Hersey 1963)

at regular intervals and the subsequent echoes from the ocean floor and subsurface reflectors are detected by a nearby line of hydrophones towed behind the source. See also →refraction, seismic instruments, and →seismic refraction shooting.

refraction the changing of the direction of propagating waves due to transversal differences in wave velocity. Refraction occurs in →optics, →acoustics and in the propagation of →gravity waves. Refraction is described by →Snell's law.

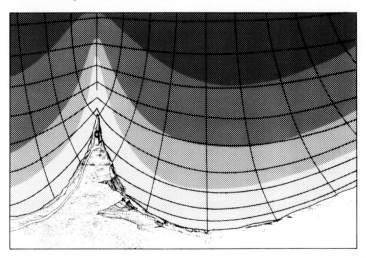

Refraction. (After Bascon 1959)

refraction shooting (seismic) (geol.) a seismic technique, based on the travel times of waves that are deflected upon entering a layer with a higher seismic velocity, move parallel to the layer contact and re-emerge at the surface again. The travel times are measured as a function of distance between shotpoint and receiver. At sea, explosives are set off at regular intervals from a moving ship and the echoes are recorded by a hydrophone on a →buoy (Sonobuoy) or a stationary ship, as the distance between source and hydrophone increases. See also →seismic instruments, seismic survey, reflection shooting, and →refraction (phys.).

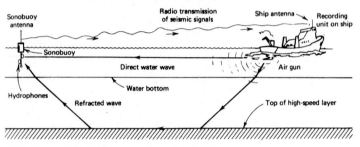

Refraction shooting (seismic). (After Dobrin 1976)

regenerated production see →new production.

regeneration the restoration by an organism of tissues or organs which have been removed (usually by predators).

regression (1) retreat of the sea from land areas by fall of sea level or uplift of land, and the consequent geographical evidence of such withdrawal, e.g., the enlargement of the area of deltaic depositions. Regressions may be worldwide as the result of global sea-level fall (see →eustasy) or locally caused by uplift of the land mass involved. Compare →transgression, →coast; (2) (stat.) relation between two quantities, established by means of statistical correlation between paired values of these quantities.

Reineck box sampler/corer the first type of →box corer invented by Reineck in 1958 to collect undisturbed samples of the sediment surface.

relict sediment sediment deposited in a previous period and in equilibrium with the environment at that time, but unrelated to its present environment even if it remains unburied by later sediment. Large parts of the North Sea contain relict sediments, deposited during the →Pleistocene and Early →Holocene.

remote sensing in general this is the technique of obtaining information on distant objects by electromagnetic (→optical or radar) or →acoustic signals. Distinction can be made between active remote sensing where the returning echo of an artificially generated signal (e.g., radar or echosounding) is measured and passive remote sensing where naturally occurring signals such as visible light or infra-red radiation is recorded. Remote sensing instruments can be mounted in satellites (see →satellite oceanography), in aircraft, ships, moorings or coastal stations. The advantage of remote sensing is the possibility to obtain detailed information over wide areas or extended periods. The problem, however, is that the relevant physical or biological parameters are not measured directly. Therefore generally a combination of physical models, empirical algorithms and mathematical and statistical techniques is needed to translate an observed signal into a quantitative estimate of the relevant parameter, and so-called ground truth or →sea truth (collection of field data) is necessary for validation.

reproduction the process by which living organisms multiply. In asexual reproduction the organisms multiply by simple division (algae, fungi, →protists, coelenterates, etc.). Continuous division can lead to formation of a large number of small identical "spores" (spore formation) developing into identical individuals. In sexual reproduction haploid gametes are formed in sexual organs, in higher plants and animals usually a large number of small "male" gametes or spermatozoa and a smaller number of large "female" gametes or eggs (ova). When eggs and sperm are mature they are released in the spawning process and →fertilization takes place when a haploid egg melts together with a haploid spermatozoid into a diploid zygote or fertilized egg. In higher plants and animals reproduction is usually a seasonal process with an annual cycle in growth of sexual individuals (e.g., →coelenterates) or reproductive organs (gonads), a limited spawning period and regression of the gonads after spawning. In lower latitudes and in smaller organisms (→copepods, nematodes) reproduction can take place more often, leading to shorter reproductive cycles. The timing of reproduction, the number of eggs produced, and all other behavioral aspects of reproduction together form the reproductive strategy of a species, aimed at achieving, under the prevailing environmental conditions, a maximal reproductive success (i.e., producing the maximal number of successful offspring in a life-time). Reproductive strategies, closely related with life strategies of species, are basically solutions to optimalization problems of energy expenditure: how to deal with the available energy in a life-time to reach maximal reproductive success.

Reptilia, marine (lizards, snakes, crocodiles, tortoises) a class of the subphylum →Vertebrata, differing from amphibians in the production of amniotic eggs, which can be laid on land because they are protected against drying and injury by a series of membranes, fluids, and a shell. This enabled reptiles to become fully land animals, some of them returned to the sea. All reptiles are →cold-blooded and as such best adapted to live in the (sub)tropics. A single lizard species, the marine iguana of the Galapagos

Islands, marine in its dependence on marine algae for food, was first
reported on by Darwin (1839). Some crocodiles and alligators frequent
estuarine parts of rivers, particularly →mangrove swamps, and are
sporadically seen at sea (some reached the Cocos-Keeling atoll alive, 600
miles of ocean travel, but did not establish themselves there). Within the
family of →sea snakes (Hydrophiidae) some genera come ashore to lay
eggs, others are →viviparous and completely marine. Like all snakes they
are strictly carnivorous and they use a venom to paralyze their prey. This
venom is toxic to man. Marine →turtles are tortoises adapted to life in the
sea by the possession of limbs that are flat flippers. Only seven species are
known. Eggs are deposited in holes dug in sandy beaches. The breeding
areas on land are often widely separated from the feeding area. Some Green
turtles feed on →sea-grass meadows off the Brazilian coast and migrate
2000 km to Ascension Island for egg-laying as found by →tagging. They
must have a remarkable navigation capacity to return to this small island
where they were born. Turtles are sometimes observed outside their normal
distribution area. Between 1950 and 1966, 61 (probably five species) were
observed in UK waters. During hot summers the area in which they can
survive may be enlarged: most specimens in UK and Norwegian waters are
sighted in June to November. There are also records from the Dutch coast.
Famous fossil marine reptiles (*Ichthyosaurus, Mosasaurus*) lived in the
→Mesozoic when reptiles were the dominant vertebrates.

research vessels specially designed or converted vessels for carrying out
marine research for physical, chemical, biological, and geological oceanographic studies. Basic instruments on board are deep-sea winches, sampling
equipment, and laboratories for immediate investigation of the samples or
preparation for storage. A modern development of marine research vessels
is the application of multipurpose containerization for laboratories,
winches, workshops, and supply units.

residence times (1) factors that characterize the general behavior of elements
in the oceans. Residence time (T) is defined by the ratio of the total amount
of material present in solution and suspension (A) in a basin (or all oceans),
and the amount introduced or precipitated (dA/dt). Conservative elements
(Na, Mg, Ca, K and Sr) have a very long T, for instance over 10^7 years for
all oceans together, while nonconservative elements have much smaller T
values, such as: Al (100 y), Pb (2000 y) and Mn (140 y). High T values
correlate with high solubility, while low T values reflect tendencies to
precipitation; (2) term used in hydrography to denote the average time a
water molecule stays in a certain sea area (e.g., in an estuary or in a shelf
area, where the water is continuously replaced by river flow or exchange
with the open sea). See also →time scales.

residual current in a sea area where the tide dominates, the effect of other
dynamic factors causing current (wind, pressure gradients) only becomes

manifest when the tidal currents are eliminated from the observations by mathematical filtering techniques. The current that remains after this elimination is called the residual current. It appears that the tides themselves, due to nonlinear dynamic behavior, may also give rise to constant, permanent currents that are part of the residual current thus defined. This tidal contribution is known as the tidal residual.

respiration (1) the oxidative degradation of nutrients (glucose, protein, fat) through metabolic reactions within the cells, resulting in the production of carbon dioxide, water and energy; (2) the exchange of gases between the cells of an organism and the external environment, including external respiration, transport by the bloodstream and internal respiration. In small aquatic animals oxygen may diffuse directly into the cells, while carbon dioxide within the cells diffuses out; in larger animals special respiratory systems have developed (gills in invertebrates and lower vertebrates, lungs in higher terrestrial vertebrates).

respiratory quotient (R.Q.) the ratio of carbon dioxide emitted and the amount of oxygen taken in. The respiratory quotient is a quantitative measure the type of food being used by the body. From the respiratory quotient one can determine whether carbohydrates (R.Q. = 1), fats (R.Q. = 0.7) or proteins (R.Q. = 0.8) are being used for respiratory needs. Usually the R.Q. value is around 0.85; at starvation the value will drop to 0.7, as the body uses its fatty deposits. At prolonged starvation the respiratory quotient may rise again, indicating the utilization of proteins from muscle tissue (life-threatening process).

resting stages (resting cells, -spores, -eggs, cysts) are specialized eggs or cysts (enclosing the entire organism) formed in a number of groups of small organisms under unfavorable conditions (of light, temperature, ionic conditions, presence of poison, poor nutrient conditions, etc.). Resting stages are resistant to such unfavorable conditions because of heavy cell walls, lowered metabolic rate (dormancy), and reduced exchange with the surrounding medium, and sink to the bottom to await more favorable circumstances. Each of the algal classes typically represented in marine and freshwater plankton (→Bacillariophyceae, →Dinophyceae, →Eugleno- phyceae, →Cryptophyceae, →Cyanophyceae, →Chlorophyceae, →Prymnesiophyceae) is characterized by one or several types of resting stages known as akinetes, statospores, cysts, etc. Resting stages also occur in a number of animal groups. They are best known in freshwater organisms (where desiccation is a major problem): statoblasts in →bryozoans, gemmulae in sponges, ephippian eggs in →cladocerans, diapause eggs and cysts in copepods, cysts in →nematodes, →rhizopodes, tardigrades and →polychaetes. In the sea resting stages are now known in →tintinnids, →copepods (diapause eggs), →cladocerans, →bacteria, and →fungi.

reticulum in protozoa, the pseudopodia (flowing extensions of the body used for locomotion) can be interconnected, in which case they form a network or "reticulum".

reversing thermometer mercury thermometer for measuring the temperature under water. The principle is that the indication of the thermometer at a given depth should be fixed, after which the thermometer is taken up again without change through overlying water layers with different temperatures, so that the indication can be read aboard. This fixation of the indication is realized by reversing the thermometer, which is constructed in such a way that the connection between the mercury in the capillary and the reservoir is interrupted thereby. By attaching the thermometer to a →Nansen bottle or some other sampler with a reversing frame a simultaneous observation of temperature and salinity can be obtained. The thermometer can be enclosed within a glass tube to protect it against pressure effects. Unprotected thermometers are used together with a protected one, and the difference between the real temperature and that with an additional pressure effect can

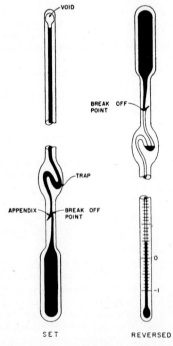

Reversing thermometer. (After Landolt-Börnstein 1968)

be used to determine the pressure at which the reversing took place, to check the depth of the observation. The temperature can be read with these instruments to about 0.01 °C. See also →potential temperature.

Reynolds number (Re) named after O. Reynolds, 1842–1912, is a nondimensional number occurring in →similarity theory, giving the ratio between inertial forces and the viscous forces in a current with a typical length scale L and a typical current speed U. It equals: $Re = UL/v$, where v is the kinematic →viscosity of the water. The transition from a laminar current to →turbulence occurs above a critical value of Re.

Reynolds stress the internal →stress due to turbulent exchange of momentum between the layers of a fluid.

Rheic Ocean (syn. Proto-Tethys) the ocean which separated the Baltic (N. Europe) and Laurentian (N. America and Greenland) →plates from the southern continents and S. Europe. The Rheic Ocean became closed during the Late →Carboniferous.

rheotaxis the movement of an organism directed by a water current. It is difficult to imagine how pelagic animals sense the direction of a current. Benthic animals may move against a current towards food, but in such a case it is more likely that →chemotaxis is involved. Good examples of rheotaxis are therefore not available from the marine environment.

rheotropism growth as a response to currents, can be found in sessile organisms, e.g., transversal rheotropism in hydroids, →bryozoans and horny corals, whose colonies grow perpendicular to the main direction of currents.

Rhizocephala order of the class →Cirripedia (subphylum →Crustacea, phylum →Arthropoda). Parasitic barnacles, almost exclusively on decapod crustaceans. They have no alimentary canal, and in the adult phase neither appendages nor segmentation. As larvae they attach by an antennule, and as adults they are fastened to the host by a stalk from which roots proceed into the host's tissues. e.g., *Sacculina*, parasite of the shore crab.

Rhizomastigina order of the subphylum →Sarcodina (phylum Sarcomastigophora, see →Protozoa), with one or two →flagella, and the whole surface of the body permanently ameboid.

Rhizopoda see →Sarcodina.

Rhodophyceae (red algae) class of the division Rhodophyta (see →algae). The multicellular members of this class range in size from 1 mm to a few dm. Rhodophyceae contain the pigments phycocyanin, phycoerythrin and lutein, but no →chlorophyll other than chlorophyll-a. They store starch outside their plastids. →Zooids are completely absent in rhodophycean life cycles. Red seaweeds live attached to rocks or sometimes on brown seaweeds (→Phaeophyceae). Several genera produce the raw material for agar-agar or carrageen. *Porphyra* is cultured for human food in Japanese estuaries. *Lithothamnion* and some other genera encrust rocks and coral reefs with their calcium carbonate reinforced bodies, or form loose nodules on the seabed. Such nodules are harvested locally as soil fertilizer. Few fossils can be assigned unequivocally to the Rhodophyceae, except the calcareous species.

Rhodophyta see →Rhodophyceae.

Rhynchocoela see →Nemertea.

rhythm, periodicity in a biological context, rhythms are physiological or behavioral changes, recurrent on a regular time scale such as days or lunar months. One cycle of changes is called a period. See also →biological clocks and →lunar cycles.

ribonucleic acid (RNA) a polynucleotide that governs protein synthesis in a cell, existing in a variety of forms serving particular functions, such as messenger RNA, ribosomal RNA, transfer RNA.

Richardson number *Ri* named after the British physicist L. F. Richardson is a non-dimensional number that gives the turbulent state of a stratified flow. If *Ri* is large, →turbulence is suppressed by →stability. Different types of Richardson numbers are used. The gradient Richardson number gives the relation between stabilizing →buoyancy and destabilizing →shear. The flux Richardson number gives the relation between energy loss by buoyancy and energy production by shear. They differ by a factor that is the relation between the coefficients for the turbulent diffusion of mass and momentum. If the flow is characterized by a typical vertical length scale L, velocity U and a reduced →gravity g', the Richardson number can be approximated by the bulk Richardson number $Ri = g' L/U^2$. In modeling stratified flow this number should be equal in model and prototype.

richness see →diversity.

rift topographic depression of regional or global extent formed by appreciable displacement along normal faults with roughly parallel strikes, and associated with earthquake and commonly volcanic activity. Rifts are an expression of tension in the earth's crust and may be precursors of →mid-oceanic ridges. e.g., the Gulf of Aden, East-African rift system.

rigid lid approximation when the deformation of the surface (for instance by waves) has no influence on the approximation of the dynamical equations, the ocean can be considered to be covered with a rigid lid. A criterion for this approximation is that the length scale of motion is far less than the Rossby →deformation radius. When the rigid lid approximation is valid, the variation of potential →vorticity, due to a deforming surface can be neglected.

rip current see →wave-induced currents.

ripple (1) small ridge in a ripple field formed on top of a layer of sediment
(most often sand) as the result of oscillating or streaming water; (2) a general
term for all sand waves with shapes similar to small-scale ripples, regardless
of their size. Types of small-scale ripples are current ripples (asymmetric)
and wave ripples (symmetric). See also →cross-bedding.

a

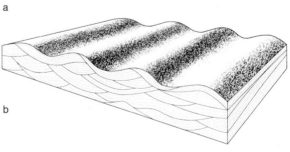

b

Ripple. (Reineck and Singh 1973). *a* Asymmetrical ripples. Land towards the
right. *b* Internal structure of asymmetrical wave ripple and resulting →cross
bedding

rise see →continental slope.

river run-off the part of precipitation on land that appears in surface streams. It does not include flow through artificial diversions, storage basins or other constructions.

river water the excess precipitation over evaporation from the continents carries $700\,000$ m^3 s^{-1} river water, loaded with small amounts of inorganic particulate and inorganic dissolved material (salts), to the oceans and seas. This particulate material, $8\ 10^9$ ton y^{-1} is produced by erosion and weathering of rocks, and is mainly deposited in estuaries and near-coastal regions. The soluble material from weathering amounts to about 90 g m^{-3} and consists of HCO_3^-, Ca^{2+}, SiO_2, SO_4^{2-}, Cl^-, Na^+, Mg^{2+}, and K^+.

RMT Rectangular Midwater Trawl, see →nets.

RNA see →ribonucleic acid

rocky shore zonation the presence of a distinctive vertical zonation pattern of communities on rocky shores, each community characterized by different dominant species. This pattern is established under the influence of the tidal regime and local topography and depends mainly on the various grades of exposure to desiccation intertidal organisms can withstand.

rosette sampler see →seawater sampler.

Rossby number (Ro), named after the Swedish meteorologist C.-G. Rossby (1898–1957), gives the ratio between the inertial forces of a current with a typical speed U over a typical length scale L, and the →Coriolis force. It equals $Ro = U/fL$, where f is the →Coriolis parameter. If Ro is small the Coriolis effect is more important than the inertia, a condition that should apply for →geostrophic balance.

Rossby wave (or planetary wave) is a north-south oscillation of zonally moving water elements under the influence of the interaction between the inertia and the latitudinal change of the →Coriolis parameter. The frequency of these waves is much lower than this parameter. This yields, in the ocean, time scales in the order of a few months and wavelengths of a few hundred kilometers. In a Rossby wave total →vorticity is conserved, with a continuous conversion of relative into planetary vorticity and vice versa.

rotary drilling the chief method of →drilling deep wells, esp. for oil and gas. A drill bit at the end of a rotating drill pipe grinds a hole in hard rock or tears into soft sediment. A muddy fluid is pumped down through the drill pipe, passes out through the bit and returns to the surface in the annular space between the wall of the hole and the pipe. The drilling fluid brings the cuttings to the surface, cools the bit and prevents the walls of the well from caving. Oil and gas exploitation in continental shelf areas involves such drilling, where besides the spoiling oil in the environment, the drilling fluid contaminates the surrounding of drilling platforms. Under optimal conditions, disturbances of the →benthic system can be restricted to a 500- to 1000-m-wide area around the platform.

Rotifera (rotifers) phylum of minute marine and freshwater animals which swim and feed by means of ciliated bands. Formerly confused with ciliated →Protozoa, who are, however, unicellular organisms contrary to the multicellular rotifers. They are small bilaterally symmetrical, unsegmented and have typically a ciliated trochal disk for locomotion and food collection, and a complete alimentary canal with anterior mouth and posterior anus, and a muscular →pharynx with specialized jaws. They have an excretory system with →flame cells joining the hind gut to form a cloaca. There is no blood system and no respiratory organ, and a very simple nervous system. There is a large perivisceral cavity which is not a →coelom. The sexes are separate, but males usually degenerate or are absent and →parthenogenesis is common. The phylum belongs together with →Nematoda and →Gastrotricha to the →Aschelminthes.

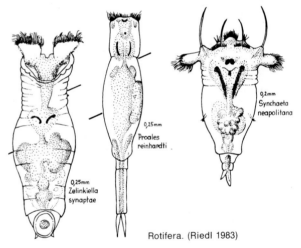

Rotifera. (Riedl 1983)

r-strategy a →life strategy which allows a species to deal with the vicissitudes of climate and food supply by responding to suitable conditions with a high rate of reproduction. The r refers to the intrinsic rate of natural increase in the →logistic equation $dN/dt = rN(1 - N/K)$, with N the population density and K the carrying capacity of the environment, as determined by food, space, predation, or other things. R-strategists are continually colonizing habitats of a temporary nature; as such, their strategy is basically opportunistic, with dispersal being a major component of their population process. At the extreme, when colonizing an ecological vacuum and thus without density effects or competition, the optimal strategy is to put all possible matter and energy into reproduction, with the smallest practicable amount into each individual and to produce as many offspring as possible. R-selection thus leads to high productivity.

S

Sagitta arrow worm, see →Chaetognatha.

salinity measure of the concentration of salt in seawater. In order to obtain accurate comparable figures, an exact definition is necessary. The first definition (1889) was based upon the determination of the mass of all dissolved salts in one kilogram of seawater by means of evaporation. This approach is not practical for routine measurements and hence was replaced by a relation based on the →chlorinity titration, which, thanks to the constant composition of →seawater, was sufficiently close to the original definition. A good correlation exists with chlorinity (the amount of chloride in grams per kg seawater): $S = 1.805 \, Cl + 0.03$. However, with the possibility of measuring the salt content of seawater as conductivity, as accuracies increased, a new definition became necessary. In 1978 the "practical salinity" was introduced, based upon the conductivity of seawater. The (practical) salinity (S) is given by a nondimensional figure, in the ocean within the range $S = 30$ and $S = 40$. In older publications, salinity (related to the older definition) is given in permilles. In most cases, where the highest accuracy is not required, old and new figures may be used together indiscriminately.

salinity crisis marine biotic crisis due to excessive salinity. Salinity substantially greater than that of normal seawater may occur when the exchange of water with the open ocean is poor, whereas the supply of salts (e.g., by rivers) continues. Moreover, an increase in water temperature or upwelling of hydrothermal water may lead to the dissolution of preexisting evaporites. The so-called →hypersaline basins are the sites of extensive evaporite formations. The most spectacular development of evaporites in a major ocean basin occurred in the western Mediterranean at the end of the →Miocene (Messinian), with evaporite thicknesses of up to 2 to 3 km in places. At that time the Mediterranean is thought to have been completely separated from the Atlantic, and the event seems to have caused an almost complete absence of life.

salinometer an instrument which can estimate →salinity from a physical property of seawater. Fundamentally physical properties such as →density, refractive index, sound velocity and electrical conductivity can be used. Since 1960 electrical conductivity has become used widely after the development of an accurate, fast salinometer by Brown and Hamon. Their inductively coupled salinometer is electrode-less, has a temperature compensation circuit and permits an accuracy of 0.003 S with proper standardization. The high temperature coefficient of electrical conductivity of seawater (3%) demands either a very good and stable temperature compensation or a measurement in a thermostat of high, long-term stability (better than 1 mK). This last principle is applied in modern four-electrode flow-through salinometers which have a 0.001 S precision.

salps see →Thaliacea.

salt marsh (syn. tidal marsh) →tidal flat regularly or intermittently flooded by the tide, fringes the intertidal zone of muddy and sandy coasts of →estuaries and protected shores in temperate and cold latitudes. Only a few macrophyte species have acquired the necessary salt tolerance to grow here. A succession of species occurs in zones from low to high, such zonation being mainly related to the height of the marsh relative to the average water level. Most of the typical genera occur worldwide: *Salicornia, Spartina, Suaeda, Limonium, Scirpus, Juncus.* The lower limit lies at about mid-tide level for the US east coast, but higher (round neap high-tide level) for European coasts. The upper limit is generally controlled by the level reached by spring tides. In the tropics similar areas may be covered by shrubs and trees and is then called mangrove swamp or mangal.

salt wedge when freshwater and seawater meet, as is the case in →estuaries, the two water masses with different densities do not in all cases mix completely, depending on the available mixing energy from wind, tide, and current. When the two water masses can be recognized in the mixing zone by their density (salinity) difference, the inward flowing tongue of seawater is referred to as salt wedge.

sampling strategy especially in marine sciences, describing the physical, chemical, and biological nature of the oceans, and studying the processes in the marine system, the subject is far too large to allow any other approach than to study relatively minute samples taken from the whole. Even the collection of a relatively small number of such samples, be it pieces of seabed, volumes of water, number of organisms, or automatic registration of some characteristic somewhere over some time period (which is also a sample), involves expensive operations including the use of specially equipped research vessels. Hence, special attention is given to the details of sampling: the choice of proper sampling tools, positions, times, frequencies, etc. In general all aimed at obtaining maximal appropriate information in return for the available funds. The planning of optimal sampling strategies finds strong support in →statistics.

sand particles of sediment in the 64 µm to 2 mm size fraction. It is usually assumed to be of siliceous composition unless otherwise qualified (e.g., "coral sand"). See also →grain size.

sandspit see →spit.

sand wave see →megaripple.

sapropel aquatic →ooze or sludge composed of plant remains, desintegrated in an →anaerobic environment in swamps or on the bottom of shallow lakes and seas. It contains more →hydrocarbon than peat. It may be a source material natural gas.

saprotrophs (syn. saprophages) →heterotrophic organisms, chiefly bacteria and fungi, which feed on dead organic matter.

Sarcodina (syn. Rhizopoda) subphylum of the phylum →Sarcomastigophora, →Protozoa which in the principal phase are ameboid, without →flagella. They move by means of pseudopodia. Usually they are not parasitic, have no meganucleus, and, though they may have a phase of sporulation, do not form large numbers of →spores after →syngamy.

satellite oceanography is →remote sensing of the ocean by means of satellite-borne instruments. By means of passive remote sensing in the infrared wave bands the sea surface temperature distribution is measured routinely. By measuring the reflectance of visible light, the distribution of →chlorophyll, →suspended matter and →Gelbstoff in the surface waters can be obtained. Active remote sensing in the microwave bands is used to measure wave heights and wave spectra, surface winds, and also the height of the sea surface. For all these remote-sensing techniques from satellites an accurate correction for the atmospheric influence on the radiation is needed. However, the presence of clouds is prohibitive for the use of visible light and infrared radiation in satellite oceanography. The satellites used are in part operational weather satellites, but satellites are also launched with dedicated oceanographic missions. Both geostationary and polar-orbit satellites have oceanographic applications.

scanning electron microscopy (SEM) is used for the study of very small particles. It can reach a higher resolution than the light microscope, because it uses an image-forming radiation with much shorter wavelength than visible light. With the SEM, particles of ca. 6 nm can be distinguished. With a narrow electron beam the surface of an object is scanned and after interaction with the material, low-energy electrons escape from the sample. These secondary electrons are used to construct an image which depicts the sample surface with mainly topographic information. Other electrons and photons escaping from deeper layers in the sample can – after each has been collected with specific detectors – give information on the chemical composition of the sample. Material to be studied with the SEM ought to meet two conditions: the sample should be dehydrated to avoid implosion in the low-pressure environment of the electron-specimen collision. As a consequence organisms cannot be studied →in vivo. Secondly, nonconductive material should be coated with a thin conductive (gold or carbon) layer to avoid image-disturbing electric charging.

Scaphopoda tooth or tusk shells, class of the phylum →Mollusca. Burrowing mollusks with a typical conical or cylindrical shell, which is open at both ends. There are no gills. The animals live head down in the sediment and feed on microscopic organisms.

scattering the changing of the direction of propagation or phase of a wave, due to small-scale variation of the properties of the medium. Scattering occurs in →optics, →acoustics and in water waves. As a criterion for the scattering properties of the medium the so-called WKB (Wentzel, Kramers, Brillouin) approximation is used, stating that the scattering can be neglected if the properties vary at length scales much larger than the wavelength.

scattering layer a concentration of marine organisms in a well-defined depth interval that scatter sound waves from an echosounder. The sound scattering is largely due to the gas-filled swim bladders (see →gas bladder) of mesopelagic fish, e.g., lanternfish (Myctophidae) and Stomiatidae. In the open ocean, these layers are very common and are called deep scattering layers (DSL). Up to five or more layers can be found down to a depth of about 1000 m. They can be more or less stationary, but others undergo daily vertical movements (→vertical migrations). In general, deep scattering layers are closer to the surface at night than during the day, due to vertical migration of the organisms.

scavengers animals feeding on dead animals.

schooling the behavior of forming aggregations with a definite mutual attraction between individuals. Grouping of individuals into aggregations is common throughout the animal kingdom. They may be more or less permanent groupings often closely knit and well coordinated or temporary aggregations dispersing in relation to fluctuating environmental conditions. The first type reaches its highest development in the sea in the schooling of fish, →Crustacea, →Cephalopoda and mammals such as porpoises. The advantage in forming such well-coordinated groups is probably to be found in an extended perception of the environment, including in some cases a protection against predators.

Scleractinia see →Madreporaria.

Sclerospongiae see →Porifera.

Scyphozoa (syn. Scyphomedusae) class of the phylum →Cnidaria. Common jellyfishes. These are cnidarians in which the free-swimming and relatively large generative stage (scyphomedusa) dominates, though they also have a sessile polyp stage on the seabed. For details on →polyp and →medusa see →Cnidaria. Most Scyphozoa develop quickly during the warmer part of the year from the small →ephyrae, loosening from the polyps during →strobilation, though in colder seas the medusae of some species live several years. They have large transparent and colored umbrellas and long tentacles with many →cnidoblasts and feed on →zooplankton, fish larvae, and other small animal prey, which they catch on the tentacles. The prey material is brought to the mouth by ciliar transport along the tentacles, and

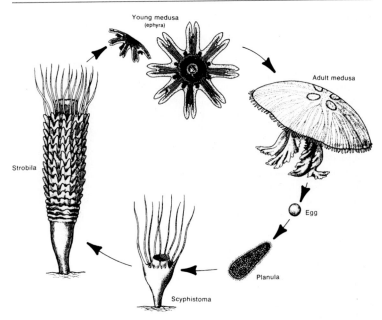

Scyphozoa. (Barnes 1987)

digested in the gastrovascular system that runs with radial branches all through the umbrella. At the end of the season, the medusae produce eggs; the ciliated planula larvae settle on the seabed to form new polyps while the medusae die and often wash ashore in great masses. Examples are species of *Aurelia, Cyanea, Chrysaora* (sea nettle), which occur all over the northern hemisphere.

sea interconnected saline waters of the world ocean with a volume of $1.4 \ 10^9$ km^3. In the special context of wave studies and wave observations the term "sea" is used for waves generated by or in equilibrium with the wind. If the wind changes or the waves travel out of their generation area, the waves are indicated by the term →swell. As the higher wave frequencies disappear rapidly outside the wind field, the crests in swell are less steep and the periods are longer and more regular.

sea anemone see →Anthozoa.

sea birds those bird species whose normal habitat and food source is the sea, whether coastal, offshore or →pelagic; →divers, grebes, pelicans, and sea ducks are generally included, although they occur at sea only part of the year.

sea cucumber see →Holothuroidea.

sea farming the man-controlled mass production of marine organisms in a sea area. A distinction can be made between intensive and extensive farming. In intensive sea farming, organisms (fish, shrimps, etc.) are kept in high densities in cages, enclosed ponds or bays with extra fertilizer or food supply. In extensive sea farming, animals are kept in the natural system (estuaries, lagoons, on rafts in the open sea) by stocking the cultured population artificially with brood collected elsewhere in the sea (as in the cultures of, e.g., mussels, oysters and clams), or reared in special hatcheries (as, e.g., the Baltic or Pacific salmon). See also →aquaculture.

sea-floor spreading see →plate tectonics.

sea-floor topography (syn. bottom topography) the general configuration of the sea floor, including its relief and man-made features. The largest-scale features are the continental shelves, reaching down to ca. 100–200 m depth, and the submarine ridges that divide the ocean floor into deep-sea basins with an average depth of ca. 4300 m. The shelf grades down towards the deep ocean along the →continental slope; at the slope basis a →continental rise and submarine fans (see →deep-sea fan) can be found. Other large-scale features are oceanic trenches [the deepest being the →Mariana Trench (11 030 m)], which can extend for several thousand kilometers. Smaller features are →abyssal hills, which cover most of the abyssal plains, and →seamounts, extinct volcanoes, which are isolated elevations rising up to 1000 m above the surrounding sea floor. Along the shelf edge and the continental slope, →submarine canyons are often conspicuous. The large-scale sea-floor topography reflects the structure of the earth's crust. The continental slope marks, in a general way, the boundary between the lighter continental crust and the heavier oceanic crust. The trenches mark subduction zones where a plate edge is moving downward below an adjacent plate. The absence of →trenches, (e.g., in the Atlantic Ocean) marks the presence of passive ocean margins. →Mid-oceanic ridges are spreading areas where new crust is formed at a rate of 1 to 6 cm y^{-1}. The sea floor is younger than ca. 150 million years, while on the continents rocks from 3.5 billion years ago have been preserved.

sea foam along coasts, especially on beaches, sea foam is a regularly occurring appeerence in spring and summer. Such foam is formed during the degradation of algal →blooms, particular of the mucus of colony-forming algae like *Phaeocystis* sp., and is blown onto the beaches at the end of the bloom period. Some foam formation occurs throughout the year, also during winter, in the absence of algal blooms.

sea grasses (syn. eel grasses) the only higher plants (angiosperms) that are adapted to living submerged in the sea (this separates them from higher plants on →salt marshes and in →mangroves). They are not true grasses, but belong to the order Helobiae, and are related to pondweeds

(*Potamogeton*). They grow in the →euphotic zone, usually in shallow subtidal waters, but in clear Mediterranean water *Posidonia* occurs down to 40 m depth and *Halophila* in the Caribbean even down to 80 m. Sea grass communities occur all over the world and have a characteristic associated fauna, many algae occurring as →epiphytes on the sea-grass leaves. Sea grasses act as sediment collectors and binders and help prevent coastal erosion. The name sea grass refers to the grass-like leaves in one of the best-known genera, *Zostera*.

sea ice any ice originating from the freezing of seawater (thus excluding →icebergs). See also →ice.

sea level as the sea level varies because of waves, tides, wind effects, seasonal variation and long-time processes, the sea level has to be defined as a mean over a certain time (often 19 years). The sea level is observed with tide gauges that already filter out higher frequencies, and the results are usually averaged over months or years. This determination gives a relative position of the sea level with respect to the measuring site, and both real sea-level changes and sinking or upheaval of the site (e.g., by compaction of sediments, by tectonic or isostatic movements of the bedrock) may have their influence. Because of density differences and currents the sea level normally does not coincide with a horizontal plane (→geoid). Sea-level slopes may be related to geostrophic motions. In the long term, the absolute sea level may change because of thermal expansion or contraction of the water, and also by: (a) melting of glaciers. During a →glacial epoch a large quantity of water is stored in →glaciers and the sea level is low; during a period of maximal glaciation the sea level is probably about 130 m lower than at present. If the climate warms, glacial melting increases and the sea level will rise; if all the present-day glaciers melted, a sea-level rise of about 65 m would occur; and by: (b) changes in the mean depth of the ocean floor. Young ocean floor is relatively warm (see →plate tectonics); cooling and shrinking with increasing age causes the ocean floor to deepen. Consequently, if the global mean age of the ocean floor is low, the mean depth of the oceans is low and the sea level is high; vice versa: if the mean ocean floor is old the sea level is low. On a regional scale, sea-level changes may also originate from isostatic movements of land masses. See also →climatic change.

seals see →marine mammals.

sea mile see →nautical mile.

seamount an isolated, steep-sloped submarine mountain, roughly circular or elliptical in plan and rising more than 900 m above the ocean floor. Most are extinct →volcanoes. See also →volcanism and →tablemount.

sea salt production of salt by evaporation of seawater (crystallization) has been practised for thousands of years and was described already by the ancient Egyptians, Greeks, and Romans. The early Scandinavians produced →brine by removing ice from freezing seawater. Crude salt was obtained by means of continuous seawater flow through shallow artificial ponds where evaporation by sun and wind followed. Almost dry, this product has a very bitter taste caused by high content of magnesium salts; it contains only about 65% →sodium chloride. The first precipitate in brine concentration is gypsum which can be removed before the crystallization of sodium chloride begins. Later on it was noticed that a much purer product could be obtained by draining a small amount of residual brine from the crude salt. Solar salt production still contributes over 40% of the total world sodium chloride production.

sea snakes see →Reptilia.

seasonal changes at increasing distance from the equator, the amount of radiation of sunlight on the earth's surface follows at an increasing degree an annual cycle, leading to darker and colder, versus lighter and warmer parts of the year with in between the transition periods (the seasons). In the sea these fluctuations are dampened considerable, due to the large heat capacity of the enormous waterbody, and due to convection: transportation of warmth by currents. Nevertheless, in the mediterranean, temperate, and boreal regions, most biological and temperature-dependent physical and chemical processes follow the annual cycle of seasons. Examples may be growth rates of algae and of the animals feeding on algae; the reproduction cycles related with energy supply to the parent and survival conditions for the brood; migrations between feeding grounds, spawning grounds, and overwintering quarters, and many other biological events in the sea that are all governed by the seasons. In tropical areas seasonal variation are related to the →monsoon.

sea star see →Asteroidea.

sea truth in order to interpret information obtained by means of →remote sensing, it is often necessary to apply a restricted number of observations made at sea of the parameters actually studied, the so-called sea truth.

sea urchin see →Echinoidea.

seawater covers 70% of the earth's surface. It is saline because it contains different dissolved ions such as those of chlorine, bromine, sulfate, sodium, potassium, magnesium, calcium. The content may vary according to the addition (by rivers or precipitation) or extraction (by evaporation or freezing of seawater) of freshwater. The oceans have an almost constant composition of salts, which only differs proportionally with the addition of freshwater. For the conservative elements sea salt is composed of 30.7% Na^+, 1.2% Ca^{2+}, and 1.1% K^+ as major cations, and 55.1% Cl^- and 7.7% SO_4^{2-} as anions. The salt content is expressed in the so-called →salinity, symbol S, that is defined on the basis of this constant composition. It is believed that the composition of seawater salt has remained constant over the last 400 million years. On the basis of S and →potential temperature, water masses can be characterized, and their →density determined. Seawater composition has similarities with that of the blood of living organisms. Hence there is little difference in osmotic pressure between seawater and blood of marine organisms. Physical properties of seawater such as →density or compressibility depend on salinity, temperature and pressure, and are calculated from these quantities with standard tables (see →thermodynamics of seawater). For accurate determination of the salinity standard seawater is made available as a reference.

seawater sampler a device for taking water samples. The usual types are lowered on a wire, and closed at a certain depth by means of an electrical or mechanical pulse. The seawater sampler that has been used most commonly in the past is the mechanically triggered →Nansen bottle. Since the introduction of electrical instruments like the →CTD recorder electro-mechanically triggered Niskin bottles have come into use. Often 12 or 24 of such bottles are mounted in a rosette sampler which can be combined with a CTD recorder.

Secchi disk a white disk of standard size (30 cm) used for estimation of the turbidity of the water. The disk is lowered and the depth at which the disk disappears from the sight of the observer is the so-called Secchi depth. Comparison with more sophisticated instruments for measuring light under water delivers empirical formulas to estimate attenuation coefficients for the visible part of the spectrum from the Secchi depth. The →euphotic zone is roughly three times Secchi depth. Devised in the 1860s by the Italian astronomer Angelo Secchi while he worked in the Mediterranean aboard the papal vessel *Immacolata Concezione*. Its strength is its simplicity and the possibility to compare data collected with the same apparatus for over more than a century. Although the method is subjective it can give a reliable estimate, if used with care. In extremely transparent waters such as occur in mid-ocean or in the Antarctic, depths down to almost 80 m can be attained, be it with disks of a larger diameter.

secondary wave see →S-wave.

sediment particulate solid material accumulated by natural processes.

sedimentary structures structures in sedimentary rocks resulting from depositional, deformational, and/or diagenetic processes. Structures may be formed either simultanously with deposition (primary sedimentary struc-tures) or subsequently (secondary sedimentary structures). A large part of the vocabulary of sedimentary structures refers to the external forms and internal structure of beds of sediments (see →cross-bedding and →graded bedding), markings on bedding planes (→ichnofossils), and penecontem-

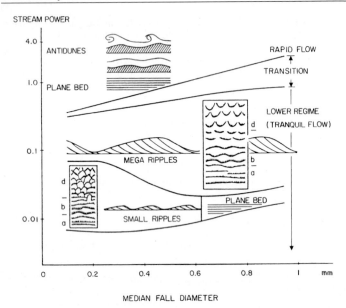

MEDIAN FALL DIAMETER

Sedimentary structures. (Reineck and Singh 1973). Bedforms and their relation to grain size and stream power. *a–d* succession of ripple from with increasing stream power

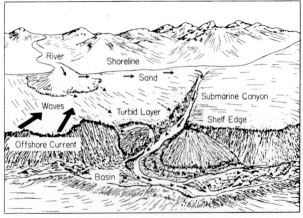

Sediment transport. (Reineck and Singh 1973). Roules of transportation of sand (*solid arrow*) and mud (*dotted arrow*) from river to deep ocean floor

poraneous (at or shortly after deposition) deformation of bedding. Bed forms depend on particle size and current velocity. Generally speaking, tranquil flow of water produces small ripples and →megaripples (more than 60 cm in length), while rapid flow produces plane beds and →antidunes. Often the distribution of bed forms is selective, so that we can distinguish sedimentary environments, such as coastal, deltaic, mudflat, slope and deep-sea environments, each having its own characteristic configuration of bed forms. →Wave ripples, for example, are restricted to shallow water environments (less than 60 m and usually from 0 to 20 m water depth), while contourites are found on the slope and →abyssal plain.

sediment classification three major types can be distinguished: (a) lithogenous sediments: →detrital products of desintegration of pre-existing rocks and of volcanic ejecta. Nomenclature based on →grain size; additional qualifiers are derived from the lithologic components (terrigenous, bioclastic, calcareous, volcanogenic, etc.) and from the structure and color of the deposits. See also →deep-sea sediments; (b) biogenous sediments: remains of organisms, mainly →carbonate, →opal and calcium fosfate (teeth, bones, →crustacean carapaces). Widespread distribution, covering approx. 55% of the sea floor. Organic sediments, although basically biogenous, are commonly treated separately. Nomenclature based on type of organism and also on chemical composition. See also →deep-sea sediments; (c) hydrogenous sediments: precipitates from seawater and interstitial water. Also products of alteration during early chemical reactions within freshly deposited sediment. Nomenclature based on origin (→evaporites) and on chemical composition.

sediment drift (or sediment ridge) large elongated ridge composed of thick piles of sediments. They are acoustically poorly stratified features, elongated parallel to the bottom-current flow. They are often hundreds of kilometers long and tens of kilometers wide, and vary from sharply peaked to slightly convex. The ridges result from the interactions of different currents or of active currents and relatively static water. They form parallel to these currents at the margin of the flow. Sediment accumulation rates are generally high, with known measured rates of at least 12 cm per 1000 years.

sediment-gravity flow mass movement of sediment induced by gravity. Four main types are distinguished on the basis of the internal support of the sediment within the flow; (a) debris flow: grains are supported by a dense mixture of water and clay with a finite cohesion; (b) fluidized sediment flow (often loosely packed sand): grains are supported by the pore fluid within the sediment (as a result of upward flow of fluid), causing sediment to move downslope (even on gentle slopes) as a traction carpet; (c) grain flow: the sediment is supported by direct grain to grain interactions; (d) the →turbidite: the sediment is supported by the turbulent flow of water (see →turbidity current).

sediment ridge see →sediment drift.

sediment transport individual particles can be transported in three different ways: (a) traction (rolling/sliding; i.e., constantly in contact with the bottom; (b) saltation (jumping; i.e., with intermittent contact); and (c) in suspension (no contact with the bottom); corresponding sediment load classifications are bedload, saltation load, and suspension load. Sub-aqueous mass movements of sediment include, for instance, →sediment-gravity flows, →turbidites, rockfalls, and slumps. Slumping is a relatively common phenomenon that occurs when a slope becomes unstable: sediment slides down the slope (usually along a curved →shear plane) as a more or less consistent mass. Besides these types there is marine glacial transport, in which (often coarse-grained) sediment is transported by ice drifts, and eolian transport, which is not strictly speaking marine transport. See also →dust. (Figure see p. 250)

seiche standing wave occurring in lakes or in semi-enclosed embayments as a result of some sudden disturbance. The topography of the basin determines the form and period of the different seiche modes that may occur. The word "seiche" was originally used for this type of oscillations in some Swiss lakes. (Figure see p. 253)

seine see →nets.

seismic instruments the instruments and sources of energy used in marine →seismic surveys are numerous and depend upon the particular purpose. The study of the deepest layers and the →Moho requires →seismic refraction shooting with explosive charges of up to 100 kg TNT. The more frequently used →seismic reflection shooting can do with smaller acoustic pulses and as conventional chemical explosives are highly impractical, a wide variety of nonexplosive sources has been developed. One of the larger energy sources used in marine seismic reflection work is the air gun, a device which functions by discharging a bubble of highly compressed air into the water. The water gun uses the discharge of compressed air to drive a piston. The piston displaces a body of water at such a speed that a vacuum cavity forms in the water. The implosion of the vacuum cavity generates the seismic pulse. The weaker sources include the boomer, the sparker, and the penetrating echosounder. The boomer generates an acoustic wave by implosion. It consists of two metal plates which are electrically forced apart, producing a low pressure region into which water rushes. Sparkers create a spark between electrodes located in the water. The heat created by the spark vaporizes the water, creating an effect equivalent to a small explosion. A penetrating echosounder is a low-frequency echosounder (often 3.5 kHz) which is able to penetrate into the upper sediment layers, usually to depths of a few tens of meters below the sea floor. The echoes of the signals are detected by hydrophones, which are sensitive to variations in pressure.

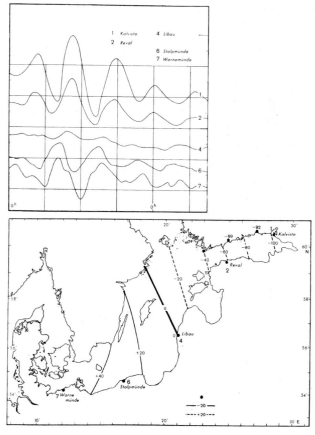

Seiche. (After Dietrich and Ulrich 1968) Sea level variations at different places in the Baltic during a seiche (upper) and amplitudes of the seiche (lower).

Several hydrophone groups are incorporated into a long cable (**streamer**) towed behind the seismic ship. Depth controllers keep the cable horizontal at a proper depth. The entire streamer may be 1 to 3 km long. See also →echosounding. (Figure see p. 254)

seismic profiling gathering of seismic data along a line. The term is usually restricted to the →seismic reflection shooting technique.

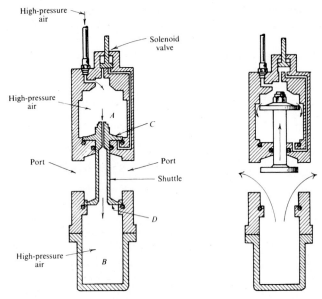

High-pressure air

Solenoid valve

High-pressure air

A

C

Port

Port

Shuttle

D

High-pressure air

B

Air gun: see seismic instruments. (After Dobrin 1976)

seismic stratigraphy study of the stratification and sedimentary features as interpreted from →seismic surveys.

seismic survey method for mapping subsurface geologic structures by initiating an acoustic pulse at a source point and observing the propagation of the →seismic waves. There are two main methods to perform a seismic survey: seismic →refraction shooting is a method not commonly used now in crustal studies, because seismic →reflection shooting gives greater continuity and takes far less effort. However, the reflection technique is not capable of reaching the deepest layers, and the study of, e.g., the →Mohorovičič discontinuity requires refraction shooting or earthquake studies.

seismic velocity rate of propagation of a →seismic wave, usually measured in km s^{-1}. The wave velocity depends on the type of wave, as well as the elastic properties and density of the Earth material through which it travels.

seismic wave general term for all elastic waves generated by earthquakes or artificial sources (see →seismic instruments). It includes body waves and surface waves. Surface waves travel along the surface of the earth and are less important seismic exploration. Body waves, however, travel through the interior of the earth and are affected by subsurface interfaces. A body wave may be either longitudinal (→P-wave) or transverse (→S-wave).

SEM see →scanning electron microscope.

semidiurnal tide a tide with high and low water twice every →lunar day. See →harmonic analysis.

sensitivity analysis see →models.

sessile a way of life exploited by →benthic organisms or organisms living attached to the substratum. The organisms are termed seston (has also a different meaning, see →seston). Sessile organisms may be found in different groups such as macroalgae, sponges, corals, bryozoans, oysters, and barnacles. There are even sessile gastropods (*Vermetus*). These organisms generally start life as →planktonic larvae and undergo a metamorphosis before they become sessile.

seston (1) (syn. →suspended matter) comprises all particles (live and dead) that are present in the water phase and have only a restricted mobility relative to the water mass in which they occur; (2) all sessile organisms; opposite of →vagon.

seta bristle or "hair" of invertebrates, produced by →epidermis.

set-up the rise of water above its "normal" elevation due to wind or waves. In tidal regions the set-up is determined with respect to the predicted tide. Prediction of the set-up is important in case of →storm surges. For shipping in shallow regions it may be important to predict the negative set-up, the set-down.

sewage see →waste and →sludge.

shading as long as there are enough nutrients in the water, algae will, under sufficient light input, show population growth. With the increasing algal density, however, the turbidity of the water also increases, limiting light penetration, and with that, growth rate of the algae. Lower in the water column, algae live in the shade of those close to the surface; this shading or self-shading affects productivity negatively.

shallow water tides it appears from →harmonic analysis that in shallow seas often not only astronomical tides occur, but also higher harmonics of the stronger constituents (for instance quarter-diurnal) or periods that are the sum of, or difference between, these strong constituents. These so-called shallow water constituents are the result of the nonlinear dynamic response of a shallow sea to the external excitation by oceanic tides.

shallow water wave see →long wave.

Shannon index see →diversity.

shear variation of current with depth. See also →friction.

shear dispersion in a current with a strong current →shear over the different layers, a dissolved substance may be spread over a large area because of the combination of differential horizontal transport in the different layers and the turbulent vertical exchange between these layers. This process has the same effect as strong horizontal →eddy diffusion, although shear dispersion often gives a more anisotropic transport.

shear instability a flow with horizontal or vertical →shear may be unstable. This means that a small perturbation in the flow may grow in size, resulting in large, wave-like or eddying motions. In vertical shear, instability is counteracted by a stable density stratification, and the criterion for the occurrence of shear instability is attained when the value of the gradient →Richardson number falls below a critical value (about 1, but depending on additional circumstances). This type of instability is also known as Kelvin-Helmholtz instability. It can result in regular disturbances in the form of overturning internal waves or →Kelvin-Helmholtz billows. The small-scale turbulence produced by this process ultimately causes an intermediate mixed layer.

shear wave see →S-wave.

shelf break see →shelf sea.

shelf sea shallow (usually less than 200 m deep) sea area bordering a continent that geologically can be considered an extension of the continent. A more or less steep slope (shelf break) forms the transition to the deep ocean. Shelf seas form less than 8 % of the total ocean area.

shellfish term used in fisheries science denoting commercial marine invertebrate species, mostly →Bivalvia, →Gastropoda, shrimps, and prawns.

shellfish farming culture of →shellfish in ponds (shrimps), in shallow basins (oysters) or in natural shallow sea areas (mussels). The idea of shellfish farming is generally to remove as many other organisms as possible from the culture substrate (which can be either natural or artificial), and stock it with settling brood collected in nature. The productivity of such artifical populations of shellfish may be high due to optimal food conditions, low mortality, and harvesting directly after the rapid juvenile growth. See also →aquaculture and →sea farming.

ship waves a moving ship generates short surface →gravity waves. If the ship (or submarine) passes through stratified water it may also generate →internal waves.

shipworm wood-boring member of the →Bivalvia which does great damage to wooden wharves, ships, etc. Example: *Teredo*.

shoal (1) shallow area in the sea; (2) aggregation of fish. See →schooling.

shoaling of waves see →wave approach.

short waves →gravity waves with a wavelength that is less than four times water depth. The →orbital motion is not influenced by the bottom. Their propagation, only depending on wave number, shows →dispersion. They are sometimes called deep-water waves.

sial granitic, gabbroic, acidic, or continental crust of the earth, plus the overlying sediments. Of dominantly silica-alumina (hence sial) composition and specific density about 2.7 g cm^{-3}. Compare →sima. See also →crust.

side-scan sonar wide-beam echosounder (see →echosounding) with sideways directed transducers for sound transmission and reception of the echoes. It produces a picture of the sea floor that can be read as an aerial photograph of the earth's surface. The technique is also used for fish detection.

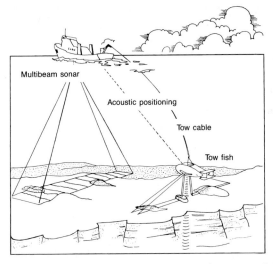

Side-scan sonar.
(After Anonymous
1983)

significant wave height is used to characterize a wave field. The significant wave height is defined as the average of the highest one-third of the waves in a record. It appears that a trained observer visually estimating the average wave height actually estimates the significant wave height by eliminating the smaller perturbations of the sea level. The significant wave height is a measure of wave energy.

silicates are salts of silicic acid H_4SiO_4. This very weak acid (pK = 9.66) occurs in all natural waters mainly as the undissociated molecule, which has a solubility of about 900 mmol m^{-3} at 0 °C and of 1500 mmol m^{-3} at 200 °C. Alternative names for silicic acid are dissolved silica and reactive silicate. The anion has a tetrahedral structure; it easily polymerizes in one, two or three directions. The three-dimensional polymer is called silica; when amorphous it contains water, the stable form of silica is quartz. Especially Al may substitute for Si in silicates. This property and the ease of polymerization explain the many different types of silicates present in rocks and clay minerals.

siliceous ooze see →ooze.

silicoflagellates belong to the →Chrysophyceae and are silica-secreting marine microplanktonic flagellates (early →Cretaceous to present), cosmopolitan, →autotrophic, with single →flagellum, unicellular. Useful in →biostratigraphy and paleoclimatology (esp. late →Cenozoic), especially in high latitudes. See also →microfossil.

silicon (Si) is, after →oxygen, the most abundant element in the earth crust. It always occurs in compounds, mainly with oxygen (see →silicates). There are three stable →isotopes, ^{28}Si (92.21 %), ^{29}Si (4.7 %), and ^{30}Si (3.09 %). The beta emitter ^{32}Si is formed by interaction of high-energy cosmic rays with argon in the atmosphere. This isotope has found some use in the dating of sediments and in circulation studies of groundwater, lakes, and oceans. The activity is extremely low and only sediments with high amounts of biogenic silica (SiO_2) can be dated. Artificially prepared isotopes have a very short half-life (^{31}Si, 2.62 h), which limits their use to tracer studies of short duration. Also mixtures of different isotopic composition (artificially prepared) are used in tracer studies.

sill lowest point on a submarine ridge or saddle at a relatively shallow depth, separating a basin from an adjacent sea or another basin. See →sea-floor topography.

silt see →grain size.

Silurian see →geological time scale.

sima basaltic, peridotitic, basic or oceanic crust of the earth, of specific density 3 to 3.3 g cm^{-3}. Name derives from silica-magnesium (hence sima) composition. Compare →sial. See also →crust.

similarity theory the use of hydraulic →models in the solution of hydrodynamic problems requires rules for the way the appropriate scaling has to be done. These rules are developed in similarity theory. In →hydrodynamics the equations are made nondimensional and in the equations thus transformed certain numbers occur that should have the same value in the original (prototype) and the model. See →Froude, →Rayleigh, →Reynolds and →Richardson number.

sinking rate of particles sinking rate or sinking velocity of particles in water is commonly defined as a steady velocity that results from the balance between the forces of gravity, →buoyancy and resistance or drag. The settling velocity of a sphere as defined by Stokes' law is the most commonly used approximation of the settling of particles. Whereas it was formerly thought that a slow but continuous rain of particles reached the sea bottom, it is now clear from sediment trap data that most particles sink rapidly and episodically as aggregates. Such aggregates may be →fecal pellets of →zooplankton or aggregates formed by other organic or inorganic processes (see →marine snow and →aggregations).

siphon opening between the fused →mantle edges of →Bivalvia, often prolonged to a tube, which can be retracted when the shell is closed. There is an inhalant and an exhalant siphon, the first leading water with oxygen and food particles into the →ventral side of the mantle cavity, the second leading the filtered water out of the mantle cavity, after it has passed the gill (see →ctenidium), which effectuates the water current by ciliary movement. In some deep-burrowing bivalves the siphons are so large that they can no longer be retracted between the shell valves.

Siphonophora order of the class →Hydrozoa (phylum →Cnidaria). Floating colonies of differentiated →polyps and →medusae that are attached to a colonial stem, and all have a special task in the colony. There is a large gas-filled medusa functioning as float and sail, there are mouthless medusae that with their pulsating swimming movements give the colonies some locomotion, feeding polyps with one long tentacle with →cnidoblasts to collect the food, and there are special polyps for the defence as well. They occur in tropical and subtropical seas; an example is *Physalia* or Portuguese man-of-war, which is very dangerous for swimmers, because of poisonous stinging cells (see →Cnidaria).

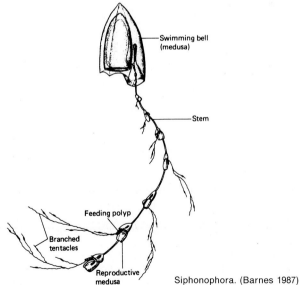

Siphonophora. (Barnes 1987)

Sipuncula small phylum of worm-like creatures, with an extensible →proboscis with spines and lobes that can be retracted into the posterior part of the body. They live on the seabed and are mostly aselective →deposit feeders. Some species live in corals or empty shells; some are borers in coralline rock.

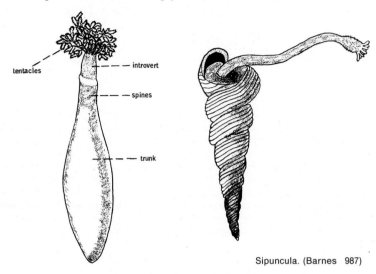

Sipuncula. (Barnes 987)

size-metabolism relationship the metabolic rate of living organisms is nonlinearly related to body size. Small animals have, per unit of weight, a higher metabolic rate than larger ones. Both in comparing animals of the same species and in comparing animals of different species, the relation can best be expressed by a power function: $M = kW^b$ (M = metabolic rate, W = body weight, k and b are constants). The constant b is always found close to 0.75, both in endothermic and exothermic species. Many hypotheses exist to explain the relative constancy of the values for k and b, none of them being completely satisfactory.

size spectra the size distributions of particulates in the oceans are often depicted in a graphical form, with particle concentration given as parts per million by volume on the ordinate and particle diameter in μm with logarithmic size increments on the abscissa. There are indications that, roughly equal concentrations of material occur at all particle sizes within the range from 1 μm to about 10^7 μm, i.e., from bacteria to whales. This implies that marine particle size spectra, with logarithmically equal intervals, are essentially flat. This pattern can only be maintained if the rate of particle production varies inversely with particle size.

slack water period in a tidal cycle when the →tidal current is near zero.

slope see →continental slope.

sludge the principal →waste produced at municipal waste-water treatment plants. It is a liquid in which up to 10 % of its weight is solids, with a bulk density of about $1.01 \, g \, cm^{-3}$, and a solid density of ca. $1.5 \, g \, cm^{-3}$. The solid fraction is a heterogeneous mixture of solids including →microorganisms, organic →detritus or aggregates, fibers, mineral grains and food residues. Also contains a film (or film fragments) floating on the surface (see →surface film).

slumping see →sediment transport.

slush see →ice.

smectite a group of expanding →clay minerals consisting of a 2:1 layer structure with the general formula: $M_1 2/3(Y_3,Y_2)_{4-6}(Si,Al)O_{20}OH_4 \cdot nH_2O$. Where $M_1 = Na$ or $1/2Ca$, $Y_2 = Mg$ or Fe^{2+} and $Y_3 = Al$ or Fe^{3+}. Deficiencies in charge in the tetrahedral and octahedral sheets are balanced by cations (mostly Ca^{2+} and Na^+) which are absorbed between the 2:1 layers. Smectites are derived from alteration of ferromagnesium minerals, calcic feldspars, volcanic glasses, hydrothermal action and dissolved Si and Fe in some deep-sea sediments. They are common in soils, sediments and sedimentary rocks. See →clay mineral and →red clay.

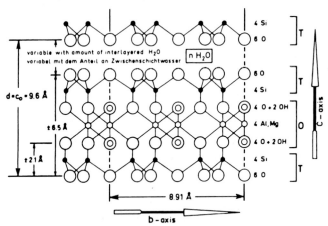

Smectite. (After Beutelspacher and van der Marel 1968)

Snell's law originally formulated for refraction of light by the Dutch mathematician W. Snell or "Snellius" (1581–1626). This law gives the change of direction of any sort of propagating wave due to differences in wave velocity.

sodium chloride (common salt, NaCl), the main constituent of the dissolved, inorganic salts of seawater. The high annual production of > 80 million tons classifies it as an important mineral. Physical constants: mol. weight 58.44, colorless cubic crystals, index of refraction 1.5442, density 2.165, melting point 801 °C, boiling point 1413 °C, solubility in water (g dm^{-3}) 357 at 0 °C and 391 at 100 °C. A quantity of 3.7 10^{16} tons is present in seawater. See also →sea salt.

sofar stands for SOund Fixing And Ranging, and is an acoustic method of locating underwater objects at large distances. It utilizes the fact that the sound speed has a local minimum at a particular depth in most oceans, which therefore is able to act as a waveguide: the sound channel, to which the energy of a sound source is refracted. As the energy decrease within such a channel is proportional to distance, rather than to distance squared (as with propagation a three dimensions) much greater signalling distances can be bridged.

solar constant solar energy falling per unit of time on a unit of surface that is perpendicular to the radiation, and that is at the mean distance of the earth from the sun (1.5 10^{11} m). The solar constant is used in formulas giving the seasonal variation of the solar radiation reaching the earth.

solarimeter see →pyranometer.

solar radiation see →light.

solar tide tide caused by the interaction of the sun and the rotating earth (see →tides). The solar tide is usually about one-third of the →lunar tide. The solar tide is indicated by S_2 and has a period of exactly 12 h.

solitary wave water waves cannot always be described by a sinusoidal variation of the elevation. Some wave phenomena have a solitary character, with each wave practically independent of the other, like waves propagating over very shallow water. After breaking near the coast the remainder of the wave can continue towards the beach as a solitary wave. In such a wave the energy dispersion by spreading mechanisms is balanced by energy concentrating mechanisms (because of nonlinear advection of momentum). As solitary waves can propagate through each other without any change in form or speed these features are often named solitons. →Internal waves often have the character of solitary waves.

soliton see →solitary wave.

sonar stands for SOund Navigation And Ranging and is a technique for locating underwater objects by recording the echoes from emitted sound signals. The principle is very similar to →echosounding, but the signals are emitted not only in a vertical beam, but in any desirable direction. Owing to this, the influence of the vertical variation in sound speed can strongly affect sound rays. By scanning over a wide zone, a kind of picture may be obtained of the sea bottom.

sorption absorption and adsorption considered jointly. These are processes occurring in solutions containing solid particles and concerning the exchange of dissolved substances with particle-bound substances. Adsorption involves the exchange of substances between a solution and the surface of particles, while absorption is the binding of substances inside particles. The kinetics of these sorption reactions are generally very different. Adsorption tends to arrive at an apparent equilibrium within a time-span of seconds to hours, while absorption may continue for hours, weeks, or months. →Desorption is the reverse process where the same kinetics as for sorption occur only for complete ion-exchange processes. Generally, desorption is much slower. In some cases sorption is apparently irreversible. Adsorption fluxes in particle surface layers of marine sedimentary particles occur with diffusion coefficients in the order of 10^{-7} to 10^{-8} cm^2 s^{-1}, while those in crystal lattices are expected to be between 10^{-12} cm^2 s^{-1} and 10^{-20} cm^2 s^{-1}.

sound in the ocean see →acoustics.

sound scanning see →echolocation.

soundings results of depth measurement that are the basis for bathymetric charts. Before 1900 wire "soundings" were the only depth data available, and the early charts were rather crude with respect to →bathymetry and restricted to shallow regions. Charts of the deep ocean could only reach their present-day accuracy by the introduction of →echosounding.

southern oscillation multi-annual (3 to 7 years) oscillation in the pressure distribution over the southern part of the Pacific, with marked effects on →El Niño, but possibly also on the weather system in other parts of the world.

sparker see →seismic instruments.

spawning the simultaneous release of eggs and sperm during reproduction. Spawning of males and females is roughly governed by their internal rhythm and ultimately synchronized by a special spawning behavior in order to ensure proper fertilization of the eggs. In many cases temperature changes, →lunar cycles or chemical substances (→pheromones) act as releasing factors for spawning (e.g., in →mollusks, annelids, coelenterates). Many species, particularly fish (e.g., salmonid fish), show spawning migrations towards favorable spawning areas, where they sometimes aggregate into spawning schools (e.g., herring).

speciation (1) (chem.) term used to distinguish between different forms or species of chemical elements in solution. For metals these species can be all forms of charged or noncharged metal combinations with the major chloride and sulfate anions, as well as with →organic matter in the form of →chelates or →complexes; (2) (biol.) the formation of new biological species from a common ancestor. When differentiation occurs as a result of growing ecological or genetic barriers within the same distribution area, speciation is called sympatric. Allopatric speciation occurs through isolation mechanisms caused by geographic separation in different areas, after which two populations of the same species evolve into different directions.

species group of related organisms or populations capable of interbreeding, mostly occupying a specific ecological →niche and designated by a common scientific name

species diversity index (see also →diversity) to express the species richness in a biological community. Most diversity indices take into account not only the number of species but also the number of individuals per species.

species interaction (types of) →intra- and →interspecific relationships: →competition, commensalism, →symbiosis, parasitism, predation, exclusion, etc.

specific heat the amount of energy required to raise the temperature of a unit of mass by one degree centigrade. The specific heat at constant pressure is higher than that at constant volume. For seawater both depend on temperature and salinity. The product of the specific heat and the mass of a body is called the heat capacity.

specific surface activity (SPA) the relative surface activity for sorption reactions. SPA is defined by the adsorption of ethylene glycol, and for ocean sediments is proportional to the empirical →base-exchange capacity.

spectral wave analysis in →wind waves an almost random pattern of crests and troughs occurs. In spectral wave analysis this pattern is assumed to be composed of a series of sinusoidal waves of varying frequency and amplitude. In a wave →spectrum obtained by processing of a wave record, the square of the amplitude (which is proportional to the →wave energy) of the composing waves is given as a function of their frequency or period. The spectrum for a fully developed wave field in deep water depends on the wind. Different mathematical functions are used as approximations of this spectrum (Pierson-Moskowitz spectrum, Joint North Sea Wave Program spectrum). For stronger winds the maximum of the spectrum shifts to lower frequencies, and the total energy density (area underneath the spectrum) increases.

spectrum a physical phenomenon such as electromagnetic radiation, where several different frequencies occur can be described by a spectral function or spectrum which represents the distribution of the energy of the phenomenon over different frequency bands. In analogy with the electromagnetic spectrum, spectral representations of any other variable (physical) parameter such as current speed or wave height can be constructed. These spectra give the distribution of the variance of the parameter over different frequency bands or over derived quantities such as period or wave number bands. Since the energy of a physical process is often proportional to the mean square, or variance, of the relevant parameter, such a variance spectrum is also called energy spectrum or power spectrum. In oceanography the spectrum is a well-known way to describe the energy distribution. The best-known examples are the wave spectrum and the turbulence spectrum. Generally such spectra have a continuous character, but for some phenomena like the tides, fixed frequencies occur and the spectrum, in extreme cases, can have the character of a line spectrum.

spermaceti liquid wax from the massive square head of the sperm whale, used in, e.g., cosmetics, candle products. Once believed to be the semen of the whale, hence "spermaceti". Probably plays a role in regulating the animal's →buoyancy during deep dives.

spheroplast is a spherical, osmotically sensitive cell derived from a bacterium by loss of some but not all of the rigid wall layer. If all the rigid wall layer has been lost, the structure is called a protoplast.

spicula (pl. spiculae) small, needle-like part of internal skeleton in, e.g., →Porifera.

spit relatively long narrow shoal, bank, or reef extending from the shore into a water body. Primarily a small point or low tongue of land consisting of sand or gravel deposited by longshore drifting, having one end attached to the mainland and the other terminating in the open water. See also →coast.

sponges see →Porifera.

spongin horny skeletal substance of certain sponges.

spore single-celled or several-celled reproductive body that becomes detached from the parent and gives rise either directly or indirectly to a new individual. Many different types usually of microscopic size, and produced in a variety of ways. Thin- or thick-walled; often produced in enormous numbers and distributed far and wide by wind, water, animals; others are resting spores, the means of survival through an unfavorable period. Occurring in some groups of plants, particularly →Fungi, Bacteria, and →Protozoa.

Sporozoa (sporozoans) see →Protozoa.

spray in coastal regions the impact of salt spray depends on the major wind directions, wind force and coastal morphology. Salt spray affects the vegetation of the spray zone (zone above the high water line), so that only salt- and wind-tolerant species will survive. Salt spray usually has a composition slightly different from that of sea salt itself, due to separation processes during bubble and droplet formation. See also →zonation.

spring tide tidal period during which the phases of the semi-diurnal →lunar and →solar tides coincide and their effects add to give a larger tidal range, with high water and low water both attaining relatively extreme values. Spring tide occurs about once every fortnight, and alternates with →neap tide after about 1 week.

stability the degree to which the effects of disturbances are compensated by restoring forces. Static stabilility in hydrodynamics is a stable situation when the fluid is at rest. When the fluid is moving the effect of additional forces has to be taken into account, to assess the dynamic stability.

stability constant (K_{st}) a term reserved for the equilibrium constant of complex formation between elements and ligands (the ions/atoms/molecules surrounding the central atom of a complex ion), either organic or inorganic. The factor is defined by the ratio of complex concentration to those of element and ligand concentrations. If more complexes are possible and the reactions are step-wise, there may be subsequent K_{st1}, K_{st2}, etc.

stable isotope see →isotope.

stagnant basin basin containing →stagnant water.

stagnant water the rate of renewal of water in most parts of the ocean is sufficiently rapid to keep the oxygen content at a high enough level to ensure normal conditions for marine life. However, in isolated marine basins the renewal may be slow or intermittent and there permanent or temporarily stagnant conditions occur, leading to oxygen deficit and the absence of most marine life forms. Examples are the deeper waters of the Black Sea, the Cariaco Trench near Venezuela and (temporarily) the Baltic Sea and some fjords.

standard deviation the square root of the →variance. See also →statistics.

standard error of mean standard deviation divided by n, where n is the number of samples taken. See also →statistics.

standard seawater seawater with a carefully checked standard composition and salinity provided by several oceanographic institutions in ampules as a reference for the determination of →salinity of →seawater samples.

standing crop (syn. standing stock) actual total →biomass in a →population or group of populations, particularly used to denote harvestable resources, but in →phytoplankton or →zooplankton indicates biomass measured per m^3 or m^2.

standing wave a wave that does not transport energy, but consists of a vertical oscillation of the sea surface with a sinusoidally varying amplitude. At distances of half the wavelength the amplitude is zero, and the water surface stays at rest. These points are called the →nodes. At either side of a node the oscillation is (vertically) in opposite phase. The water particles at the nodes are oscillating horizontally; at the antinodes, the points of maximum vertical amplitude, the horizontal movement disappears. Standing waves can be regarded as the superposition of two equal →progressive waves, travelling in opposite directions. In a (semi-)enclosed basin this can be caused by a reflection at the boundaries. A →seiche is an example of a standing wave.

starfish see →Asteroidea.

stationary phase is the period during the growth cycle of a population of microorganisms in which growth ceases.

statistics (1) mathematical methods of analyses, which are intended to aid the interpretation of data subject of a haphazard nature. Such variation is described in terms of probability →distribution. So the data are characterized by models with a systematic part and a random part, both containing only a limited number of quantities, called parameters. Assuming some probability distribution, parameters can be estimated and hypotheses concerning these parameters can be tested. To give an example: a sample containing hundreds of x_1, x_2 observations can be characterized as being bivariate normally distributed, with the means and variances of x_1 and x_2 and the correlation coefficient between x_1 and x_2 as parameters. The hypothesis that the correlation coefficient equals zero can be tested; (2) estimated parameters. Known statistics for a number of samples are are →mean (x), →variance (s^2), →standard deviation (s), →standard error of mean and the correlation coefficient. The standard error of mean is often used to describe limits to the reliability of sample means as an estimate of population means. The standard error of mean decreases as n increases, i.e., when more samples are taken the reliability of estimates of population parameters increases. However, the standard error of mean does not provide a mathematically objective criterion. Statistically computed confidence limits are easier to interpret.

statocyst organ of balance (orientation of the gravitation), consisting of a chamber containing statoliths, i.e., granules of lime, sand, etc., which stimulate receptor cells when the animal changes direction. See also →otolith.

Stefan's law this physical law, also known as the Stefan-Boltzmann law, states that a body emits an amount of radiation energy proportional to the fourth power of the absolute temperature. In this way the earth, including the ocean surface, loses to outer space energy received earlier from solar radiation. This radiation is measured by →remote sensing techniques in order to determine the temperature of the ocean surface.

steno- prefix, meaning narrow.

stenobath refers to species that can tolerate only small variations in pressure (depth).

stenoecious refers to a species that can only live in a environment in which conditions fluctuate within narrow limits. Most marine organisms are stenoecious, in contrast to estuarine species, which can usually withstand large fluctuations in environmental parameters; opposite of →euryoecious.

stenohaline refers to species that can only tolerate narrow changes in environmental salinity, and thus are unable to tolerate a wide variation in osmotic pressure of the environment. Animals of the open sea are usually stenohaline: they will not invade the brackish water of estuaries. Many brackish-water animals are the opposite, euryhaline, capable of tolerating large changes in environmental salinity. Most freshwater species are also stenohaline, but some exceptions occur with species that leave their normal habitat during the spawning period (e.g., eels, salmon, sticklebacks, etc.).

stenoionic refers to an organism that can only tolerate small variations in the ion concentration of its environment. Many marine and freshwater species are stenoionic.

stenophagic (syn. stenophagous) refers to animals with a very narrow food specialization, e.g., the ctenophore *Beroe gracilis*, feeding exclusively on another ctenophore, *Pleurobrachia pileus*. Many species prefer some diversity in their food but have the disadvantage that they cannot develop specialized feeding mechanisms by which a lot of competition can be avoided. Thus, stenophagic feeding habits may have advantages when a certain type of food is abundant and reliable.

stenotherm(ic) (syn. stenothermal) refers to species that show a distinct preference for a small range of environmental temperatures. The reason for such a preference is that by choosing a habitat with a fairly constant temperature they can attain a certain degree of stability in their metabolic rates. On land, temperatures fluctuate more rapidly and severely than in an aquatic environment (related to differences in heat capacity of water and air). Stenothermal species cannot withstand large fluctuations. In the sea many stenothermal organisms survive by means of an active habitat selection (→migration). On land organisms are forced either to tolerate temperature changes and become →eurythermal or to develop →homeo-thermal mechanisms.

Stokes drift the movement of a particle in a variable current with respect to the local mean current. This differential movement is the consequence of a transformation from the oscillating motion in the Eulerian sense to a net movement in the Lagrangian sense (see →Euler-Lagrange transformation). In harmonic gravity waves, therefore, the →orbital motion does not result in closed tracks of the water particles.

storm surge rise of the sea level above the normal tide, because of atmospheric effects, either the →wind stress or a low atmospheric pressure (inverse barometer effect). Storm surges are not only due to direct, but also to indirect effects, the latter in the form of free long waves that stem from atmospheric effects acting at some distance. Storm surges generated outside the area where they are observed are called external surges.

strain a population of cells all descended from a single cell; a clone.

strait see →channel.

stratification in the seawater column the occurrence of more or less horizontal water layers in the sea as a consequence of differences in density (i.e., in salinity or temperature). The persistence of stratification depends on the static stability of the stratified water column. The water is stable if a parcel of water, moved from its original level, experiences a gravitational force that drives it back to its original depth, and unstable if the gravitational force moves it further away (convection). The stability depends on the degree of increase of density with depth at a given location. The larger the stability, the more energy is required for vertical mixing of the water column. In →internal waves the restoring force that makes the water elements oscillate, and the maximum limiting value of internal wave frequency, the →Brunt-Väisälä frequency, depend on the stability. In the upper layers of the sea the often observed form of stratification is one marked by a sudden decrease of temperature with depth (→thermocline). Here the stability is large and vertical mixing is difficult, so a thermocline may persist for long periods.

stratigraphic correlation determination of stratigraphic relation between different rock units, based on geologic age (time of formation), lithologic characteristics, fossil content, or any other property. The term almost always denotes time correlation. See also →stratigraphy.

stratigraphic hierarchy the nomenclature of stratigraphic units depends on the type of →stratigraphy used. There is a stratigraphic hierarchy in the names arranged below in order of magnitude. Chronostratigraphical terms are time-rock terms. Bio-, litho-, and magnetostratigraphical terms are not strictly equivalent to geochronological terms and are therefore placed separately. Bio-, chemo- and magnetostratigraphic units (in the →chronostratigraphical sense) are referred to as zones, and their geochronological equivalents are chrons. Prefixes super- and sub- can be added for further subdivision.

stratigraphy study of layered rocks (sediments and volcanics), especially their sequence in time, the character of the rocks and the correlation of beds in different localities (see →stratigraphic correlation). Stratigraphy distinguishes units and boundaries based on a number of different aspects: form, distribution and lithologic composition of the beds (→lithostratigraphy), their fossil content (→biostratigraphy), and geophysical and geochemical properties, such as the succession of intervals of remanent magnetism (→magnetostratigraphy), the ratios of stable isotopes (isotope stratigraphy), the chemical composition (→chemostratigraphy), and the seismic properties (→seismic stratigraphy).

stream function a mathematical expression that describes the pattern of a non→divergent, two-dimensional flow. Lines in the horizontal plane where the stream function is constant are →streamlines giving the direction of the current. Between two streamlines flows a fixed quantity of the fluid per unit of time. The current speed is inversely proportional to the distance of the streamlines.

streamer see →seismic instruments.

streamline line in a fluid, the tangent of which is everywhere parallel to the current vector. In stationary currents it is equal to the path of the stream. In two-dimensional non→divergent flow, the streamlines are given by a →stream function.

stress (1) (biol.) any stimulus that disturbs or interferes with the normal physiological equilibrium of an organism. In →ecotoxicology the term environmental stress is sometimes used to indicate the interference of a specific ambient (→biotic or →abiotic) parameter with the normal functioning of an organism, or population. Stress is nonspecific and as such meaningless and should therefore not be used in biology science; the concept of stress originates from psychology; (2) (phys.) tangential force per unit of surface area as a result of →friction. In oceanography →wind stress over the sea and bottom stress are important factors in the ocean dynamics. Internal stresses are caused by current →shear.

strobilation see →Cnidaria.

subduction see →plate tectonics.

sublittoral zone see →zonation.

submarine canyon elongated cleft running partly across the →continental shelf and →slope. See also →deep-sea fan.

substrate is in biochemistry a compound undergoing reaction with an enzyme. In microbiology substrate is the food for the culture of the microorganisms. Substrate or substratum can also be the material, sediment, or solid object on which animals live or to which they are attached.

subtidal zone see →zonation.

succession the change in species composition with time. In plant and animal communities modification of the physical or chemical environment of the community leads to succession, e.g., the change in species composition of algae during the year, caused by the change in temperature, light, grazing, or nutrient composition due to nutrient uptake by the growing algae. Succession leads to stabilization of an ecosystem.

sulfate chemical symbol SO_4^{2-} is the predominant form of sulfur in seawater, where its concentration is 28.3 mmol l^{-1}. Sulfur in the form of sulfate is after chlorine, sodium, and magnesium the most abundant element in seawater. It can be considered a nutrient because all living organisms need it to build organic macromolecules, e.g., proteins.

sulfate reduction is the reduction of sulfate to sulfide. This process is executed by the widely distributed sulfate-reducing →bacteria. These →anaerobic bacteria use sulfate as electron acceptor in the oxidization of organic matter. During the electron transport along the electron transport system →ATP is synthetized. Such a sulfate reduction is also called dissimilative sulfate reduction. Many organisms, including higher plants, algae, fungi, and most bacteria, use sulfate as a sulfur source for biosynthesis. This is the assimilative sulfate reduction. The obligate anaerobic sulfate-reducing bacteria play an essential role in the →mineralization of organic matter in anaerobic marine ecosystems, sediments, anaerobic fjords, and basins.

sulfides compounds with S^{2-} as anion. In sediments beneath regions of high productivity, where dissolved oxygen in pore waters is quickly consumed, FeS (mackinawite or graygite) is formed from Fe (II) and H_2S which originates from sulfate by bacterial reduction. This precipitate causes the black layers found a few mm to several dm below the sediment surface. At deeper levels, the black color disappears again by transformation of FeS into FeS_2 (→pyrite) or other polysulfides which have a gray color. Sulfides are also precipitated from hydrothermal fluids near deep-sea →hydrothermal vents in →mid-oceanic ridges. They consist of mixtures of Fe, Ca, and Cu sulfides and contain also Co, Mn, Pb, and other trace constituents.

sulfur bacteria are a group of bacteria that possess the capacity of →chemolithotrophically oxidizing reduced sulfur compounds. See also →sulfur cycle).

sulfur cycle a biogeochemical cycle effected by various types of microorganisms. Assimilative and dissimilative →sulfate reductions reduce sulfate to respectively reduced sulfur in organic matter and sulfide. Different groups of →bacteria are specialized in the oxidation of sulfide. Colorless sulfur bacteria oxidize sulfide with oxygen or nitrate and phototrophic sulfur bacteria can use reduced sulfur compounds as electron donors to assimilate CO_2 in the light (anoxygenic →photosynthesis). Also some cyanobacteria are known to possess, under anaerobic circumstances, an anoxygenic photosynthesis system. In the dark these organisms use the produced elemental sulfur for →anaerobic respiration and reduce it back to hydrogen sulfide. The sulfur cycle links air, water, and sediment. A set of microbiologically mediated oxidation and reduction reactions brings about the key exchange between the available sulfate pool in the overlying water and the reservoir of precipitated sulfides (mainly ironsulfide and →pyrite) in the sediment. The natural cycle of sulfur is disturbed by large-scale burning of fossil fuel, which liberates sulfur oxides (SO_2 and SO_3) in the atmosphere. These sulfur oxides are important contributors to industrial air pollution and acid rain formation.

sun compass reaction see →light compass reaction.

super ocean see →Panthalassa.

supercritical flow current condition in which the current speed exceeds the speed of propagation of →long waves, so that disturbances cannot propagate upstream. It occurs for strong currents over shallow depths (see →Froude number) and is not a normal condition in ocean dynamics.

supralittoral zone see →zonation.

surf see →wave approach.

surf zone see →zonation.

surface film accumulation of floating organic material, such as fatty acids and →hydrocarbons, in a very thin often monomolucular surface layer (< 0.2 mm thick). They may result from the spillage of surface-active substances (oil) or from naturally occurring substances. They are very effective in damping →capillary waves. Also, particulate carbon concentrations in this microlayer are one order larger than in the subsurface, and sustain a specific flora and fauna. See →neuston.

surface microlayer interface between water and atmosphere. Since the surface microlayer is a boundary between a polar phase (water) and a nonpolar phase (air) it accumulates organic substances with nonpolar groups from the water column. The surface microlayer can be disrupted by turbulence in the water column.

surface tension tangential force resulting from molecular interactions, acting at the transition between two fluids or between a fluid and a gas. The surface tension tends to minimize the surface area of a fluid body. It is the restoring force in →capillary waves. Monomolecular layers of organic origin occurring at the →surface microlayer may influence the surface tension. Also the effect of decreasing surface tension of the sea surface on sea birds can be mentioned: at lower values seawater penetrates the fine meshwork of insulating feathers, decreasing the air retention and causing loss of insulation, ultimately lethal to the birds. See also →surface films.

surge (1) rise of the sea level above the normal tide. In the case of a large, often devastating surge caused by atmospheric forces one speaks of a →storm surge. Surges may also have an earthquake or an explosion as their origin (see →tsunami); (2) the sudden, very rapid flow of a glacier. During a surge a large volume of ice from an ice reservoir area is displaced downstream at speeds of up to several meters per hour. In the interval between surges, the ice reservoir is slowly replenished by accumulation and normal ice flow.

survivorship curve used in population biology. When a cohort (an age class in a population) is followed through time and the numbers of survivors are counted regularly, such data are compiled in a →life table and the number of survivors can be plotted (usually logarithmically) against time, giving a survivorship curve (or →mortality curve). Useful in the comparison of populations of a species in different areas.

suspended matter all particles that occur in suspension in the water column, both organic (living and dead) and inorganic. Usually defined as material left on a 0.45 μm pore-size filter. Contains the food of suspension feeders (see →filter feeders).

suspended particles see →suspended matter.

suspension feeder see →filter feeder.

suspension load total of sedimentary material that remains in suspension (i.e., afloat). Often expressed as weight or volume per unit of time.

Sverdrup relation from the balance of →vorticity it follows that for water moving to the north or south the change in planetary vorticity should be compensated. For the overall balance it is required that this balance should come from the mean curl of the wind stress. This balance is called the Sverdrup relation.

Sverdrup (unit) unit (sometimes used) for volume transport in sea currents, named after the Norwegian oceanographer H.V. Sverdrup (1888–1957). One Sverdrup equals 10^6 m^3 s^{-1}.

Sverdrup wave a long, plane, propagating wave for which the →Coriolis force on the moving water particles is not compensated by a pressure gradient as in the →Kelvin wave, but which makes the water particles follow tracks that are ellipses in the horizontal plane. For a Sverdrup wave it is necessary that the angular frequency of the wave is greater than the →Coriolis parameter. In a semi-enclosed basin Sverdrup waves may be reflected and the superposition of different waves may result in partially →standing waves, which, however, are under the influence of the Coriolis force. Such standing waves are often called Poincaré waves, although, for priority reasons, this term is sometimes used to denote the general class of propagating →plane waves.

swash see →wave approach.

S-wave (syn. secondary wave or →shear wave) type of body wave (see →seismic wave) that propagates by oscillation of the particles perpendicular to the direction of motion (shearing motion). It does not travel through liquids. Its speed is 3.0 to 4.0 km s^{-1} in the crust and 4.4 to 4.6 km s^{-1} in the upper mantle. The S stands for secondary; it is so named because it arrives later than the →P-wave.

swell sea waves that have left the area where they have been generated by the wind or that have remained after the generating wind has disappeared. See also →sea.

swim bladder see →gas bladder.

swimming speed the speed by which aquatic organisms can move in water is generally related to the size of the organism. Swimming speeds in fish are therefore often expressed in "body length per second". A distinction can be made between maximum swimming speed, maintained only for a short time ("sprint") for escape or prey catching, and cruising speed when the animal travels over long distances for hours or days. In many fish species cruising speed is approximately 1 to 2 body lengths per second, while maximum speed is approximately 4 to 10 body lengths. Some microscopic organisms, such as the ciliate *Mesodinium*, can make very fast "jumps" at an estimated velocity of more than a hundred times the body size per second, though the absolute swimming speed is small.

symbiosis the relationship between different species that live together to the advantage of one of the two or both. Types of symbiosis are commensalism (sharing food), endoecism (hiding in the host's proximity), epizoism (living attached on the surface of the host) and inquinihilism (hiding inside the host). Examples: symbiotic algae producing oxygen and →organic carbon used by the hosts such as →corals, →sea anemones and giant clams; (polychaetes living together with a hermit crab inside a shell and feeding on the its food; symbiotic bacteria, which live in the gut of, e.g., →sea urchins, and produce essential →amino-acids for the host by nitrogen fixation. Phoresis is a form of symbiosis in which one species lives in the shelter of, or is transported by, another without food relationships.

symbiotic algae see →symbiosis.

sympatric speciation see →speciation.

syncytium mass of cytoplasm, enclosed in a single continuous plasma membrane, containing many nuclei. The cytoplasm can have different forms.

synecology see →ecology.

synergism by interaction the total effect of two agents exceeds the sum of the effect of either single agent.

syngamy fusion of gametes in fertilization.

synoptic an approach in which a certain process or condition is studied or displayed simultaneously over a wide area.

syntrophy is a nutritional situation in which two or more organisms combine their →metabolic capabilities to catabolize a substance which neither can be catabolized by one of them alone.

T

tablemount (guyot) a flat-topped →seamount rising from the floor of the ocean like a volcano but planed off on top and covered by 200 m or more of water. See also →sea-floor topography and →abyssal hill.

tagging attaching labels (of which a variety of types are available) to live fish, birds, marine mammals, etc., after which the tagged and individually recognizable animals are put back in their natural environment. Under the assumption that tagged individuals and untagged behave alike, the recaptured tagged ones give information on migration, growth rates, maximum age, etc., while the ratio of tagged/untagged individuals in a catch and the known number of tagged individuals gives an indication of the population size. For recovery of tagged fish, scientists have to rely on fishermen, who are encouraged to deliver tagged fish for a financial reward. When the landings of a fish are partly or wholly reduced to fish meal, use can be made of magnetic tags, which can be recovered by magnetic separators in fish meal plants. In some cases acoustic tags (fish) or radio-tags (marine mammals, sea-turtles) are applied in behavioral studies. In such studies the changing position of an individual in the water column in relation to the tides can be followed.

Tanaidacea order of the class →Malacostraca, subphylum →Crustacea (phylum →Arthropoda). Small marine crustaceans resembling →cuma-ceans or →isopods, with a →carapace fused to the first two thoracic segments, and typically the second pair of thoracic appendages very long and suitable for grasping particles. They live in soft bottoms and collect →diatoms and other small food particles.

taxis is a movement oriented by a stimulus. See also →chemotaxis, →phototaxis and →rheotaxis.

taxonomy science of the classification of living organisms. It has two aspects, the identification and naming of each kind of organism (nomenclature) and the grouping of these according to assumed relationship (phylogeny). Modern taxonomy started with Carolus Linnaeus (1707–1778), the great biological encyclopedist who dominated 18th century biology, the age of the encyclopedists. Most influential contributions to biology are his successful introduction of consistent binomial specific nomenclature for animals and plants and his system of classification. His *Species Plantarum* (1753) and *Systema Naturae* (1758) have been internationally accepted as starting points for biological nomenclature of plants and animals. Linnaeus accepted species as given, created, and not changed since their creation; 100

years later Darwin and Wallace argued that new species originate by modification during →evolution.

tectonics study of the structural features of the earth and their causes.

tektite small, usually thumb-sized, black to greenish rounded body of silicate glass. Although tektites are of nonvolcanic origin, some may appear as solidified droplets from a great splash of molten rock. Tektites are found in groups in several widely separated areas of the earth's surface and are generally unrelated to associated geologic formations. They are believed to be of extraterrestrial origin or the product of meteorite impacts on terrestrial rocks.

telemetry distant measuring of environmental or biological variables, by applying radioelectrical techniques. The remote monitoring by a transmitter of the whereabouts of an animal is termed radio tracking and remote monitoring of its behavior or physiology (e.g., posture, heart rate, deep body temperature, levels of metabolites) is termed biotelemetry. Both techniques require the test animal to be caught at least once to equip it with a miniature transmitter, which may be implanted or attached outside, by glue or harnass. In marine species, telemetry has successfully been used in some relatively large animals (mammals, fish and birds).

Teleostei see →Pisces.

telson the hindmost segment of the arthropod abdomen. See also →Crustacea.

temperate zone temperature zone of the earth between the polar zone and the tropical zone, usually separated from the first by the 10 °C isotherm for the warmest month (roughly limit for growth of coniferous trees), and from the tropical zone by the 18 °C isotherm for the coldest month (roughly the limit of palm trees and reef corals). See also →biogeography.

temperature of seawater determines its physical properties, together with the →salinity. It is measured →in situ at different depths with instruments such as the →bathythermograph, the →CTD recorder or →reversing thermometers. In oceanography temperature is measured in degrees Celsius (°C). For some applications degrees Kelvin (K) are used, where $K = °C + 273$. Its horizontal and vertical variations are caused by →air-sea exchange of heat, mixing and advection of water of different temperature. Because of the predominant influence of temperature on the →density of seawater the temperature profile commonly shows a decrease in temperature with depth, although regionally temperature inversions may occur, where the change in salinity may counteract the effect of temperature. Also →adiabatic compression may cause a downward increase in temperature at great depths. See also →potential temperature.

temperature-salinity diagram diagram with (potential) temperature (T) and salinity (S) as coordinates, used in the analysis of oceanographic observations. Curved lines printed in the temperature-salinity (T-S) diagram give the density according to the →equation of state. Each individual observation of temperature and salinity is represented by a point in the diagram. Observations at subsequent depths give points arranged along lines that are characteristic of the water structure at a given location. See also →water mass.

Tentaculata class of subphylum →Ctenophora (phylum →Cnidaria), with retractile tentacles. Most Ctenophora belong to this group.

tephra (volcanic ejecta, ejectamenta) collective term for all fragmental materials ejected by a volcano and transported through the air; includes volcanic dust, ash, cinders, pumice, and bombs.

terrace, marine narrow coastal strip, sloping gently seaward, consisting of loose sediment or rock and veneered by a marine deposit (silt, sand, fine gravel). See also →coast.

terrestrial sediments sediments deposited on land.

terrigenous sediments sediments consisting of material eroded from the land surface. This material is liberated from igneous, metamorphic, and preexisting sedimentary rocks by chemical and physical weathering, and transported by water, wind, or ice. Most of the surface sediments in shelf seas consist of terrigenous material.

territorial sea defined by the international conferences held in The Hague in 1930, under the auspices of the League of Nations, and by →UNCLOS I, II and III as a 12-→nautical-mile zone, measured from the →baseline on which the coastal state has absolute sovereignty. Originated from the vague "cannon-shot" rule the USA and Britain had earlier settled on a 3-mile zone, which was by no means universally accepted. Scandinavian countries originally had a 4-mile limit, France 6 miles, and Czarist Russia 12 miles.

territorialism some animals have a restricted home range and defend either individually, in pairs, or in groups a specific area against conspecifics for feeding, resting, reproduction, or other activities. By such territorial behavior, which often occurs in crabs, fish and birds, population densities may be restricted, while within a territorial population →distribution tends to approach an even distribution.

Tertiary see →geological time scale.

tertiary producers/production see →production, tertiary.

testis organ which produces sperm. In →vertebrates it also produces sex hormones.

Tethys the late →Mesozoic sea which separated →Laurasia from →Gondwana. The Tethys originated from the break-up of →Pangea and was closed by the collision of Eurasia (Europe and Asia) with Africa and India during the →Neogene. See also →continental displacement.

texture (of sediment) the microgeometry of sediment, determined by the →grain size, and the shape and arrangement of grains.

Thaliacea class of the subphylum →Urochordata (phylum →Chordata). Salps, planktonic tunicates. The body wall consists of a single layer of epithelial cells, and is covered by the characteristic thick and soft tunica, similar to cartilage. The body consists of three regions, being the frontal part with the perforated →pharynx, the posterior region with the guts and other internal organs, and the post abdomen, containing the heart and reproductive organs. They are →filter feeders pumping water from the buccal siphon opening, through the sieving →pharynx, into the internal space or atrium, which empties through the atrial siphon. Thaliacea live mainly in tropical and subtropical seas.

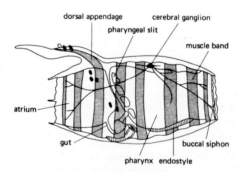

Thaliacea.
(Barnes 1987;
Hardy 1958)

thermal conductivity of seawater transfer of heat in a substance by molecular interaction. In gases and liquids the mechanism of heat transfer, by convection and →turbulence, is more important. In the sea, if convection and turbulence are suppressed by stable →stratification thermal properties are conserved over long periods and distances.

thermal expansion the increase in specific volume (see →density) with increasing temperature. In seawater the rate of increase, given by the thermal expansion coefficient, depends on the salinity. For freshwater, below 4 °C this coefficient is negative, but for normal seawater it is always positive.

thermal pollution effects due to man-made heat input in natural systems, e.g., due to input of hot cooling water. See also →pollution.

thermal spring hot-water spring, typically occuring on land, rather than on the deep-sea floor. The springs on Iceland are a prime example, Iceland being the surface expression of the Mid-Atlantic Ridge. See also →hydrothermal vents.

thermal wind relation originally in meteorology, also used in oceanography for →baroclinic flow, denoting that the rate of change of the horizontal current with depth is related to the horizontal gradient of the density field.

thermistor a temperature-sensitive conductor. The electrical resistance of a conductor is always to a certain extent dependent on temperature. Some materials have a very high temperature coefficient, mostly negative (NTC resistors), sometimes positive (PTCs). Such materials which strongly change their conductive properties in response to temperature changes can be used as temperature sensors. By measuring the electrical resistance with a Wheatstone or Scheering bridge, one obtains a signal that is (nonlinearly) related to temperature. The bridge (balance) method permits an extremely high sensitivity. The disadvantage that a thermistor produces heat (electric current is passing through a resistor), can be minimized by measuring pulsewise. Thermistors can also be used to measure tiny water movements, as their actual temperature results from the heat they produce and the rate at which this heat is dissipated to the medium, while dissipation is larger when the medium flows faster along the thermistor.

thermocline layer at some distance below the surface where the temperature rapidly decreases with depth (see →stratification). During the summer a seasonal thermocline often develops at a depth between 10 and 100 m which disappears in winter. Large ocean areas also have a permanent thermocline that is less sharp and is often defined as reaching down to the more homogeneous deep water masses. This permanent thermocline may cover a depth range from 100 m down to over 1000 m, depending on water structure and location. On quiet days a diurnal thermocline may also develop on top of the seasonal thermocline. (Figure see p. 280)

thermodynamics of seawater the relation specified between state variables such as temperature, pressure, volume, and chemical composition of seawater. As a great number of chemical substances occur in seawater, this relation is theoretically rather complex, but in practice, because of the constancy of its composition, seawater can be considered a two-component

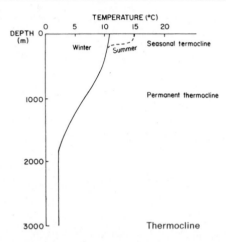

system, with →salinity as a single, alternative state variable replacing the larger set of individual chemical substances. The empirically determined international →equation of state for seawater formula was established in 1980. In the first law of thermodynamics the balance of energy of the system is described. This leads to the definition of the internal energy which increases when heat is taken up or work is done on the substance by compression (or the reverse for expansion). Enthalpy is the part of the internal energy used to change temperature or phase (vapor-water-ice). The difference of →specific heat at constant pressure or constant volume is a consequence of this first law. In →adiabatic processes enthalpy changes by compression or expansion. The second law of thermodynamics leads to the definition of →entropy.

thermohaline circulation circulation in the ocean that is driven by the density differences caused by temperature and salinity differences. Thermohaline circulation predominates over wind-driven circulation in most →estuaries and in deeper parts of the ocean.

thermophile adapted to life at high temperatures as found near →hydrothermal vents.

Thiobacteriaceae a group belonging to the →chemolithotrophic sulfur bacteria, e.g., Thiobacilli.

Thoracica order of the class →Cirripedia (subphylum →Crustacea, phylum →Arthropoda) with an alimentary canal. They have a →carapace of calcareous plates. There are six pairs of thoracic limbs and no abdominal segments. They live permanently attached, and are not parasitic. Examples: *Lepas* (goose barnacle), *Balanus* (barnacle).

thorax see →Crustacea.

thymidine assay technique to measure secondary production by bacteria. The assay measures the incorporation of radioactively labeled thymidine into →DNA, but the conversion to an overall estimate of production rates rests on assumptions that may not always be valid.

tidal component see →partial tide.

tidal current current caused by the →tides. Depending on the character of the tidal wave, the tidal current may be alternating as →flood or →ebb currents (flowing in only two opposed directions) or rotary (turning more or less gradually in a clockwise or anticlockwise direction with a current vector that describes a circle or an ellipse). In coastal sea areas tidal currents are often much stronger than the currents due to other causes.

tidal excursion range of the trajectory a water element describes during a tidal cycle under the influence of the →tidal current.

tidal flats an extensive near-horizontal, marshy or barren, sandy (sand flat) or muddy (mud flat) area lying between the levels of mean high and mean low tide. They are especially well-developed in areas sheltered against strong wave action. High parts may be flooded only during →spring tides or →storm surges. Different kinds of →algal mats may be present on tidal flats; in the tropics high tidal flats are often covered with mangrove (see →salt marsh). Sediments vary from coarse sand and gravel to fine muds. The tidal channels contain the coarsest sediment, the innermost landward flats the finest and muddiest. Often extensive ripple systems are formed on tidal flats. Suspended sediment and muddy deposits are usually concentrated on the inner flats due to a combination of tidal asymmetry, sediment consolidation during exposure, shallow depth around high tide slack water when the water covers the flats, mud particle retention by organisms (→benthic algae, salt-marsh grasses, mangrove), and shelter against waves. Tidal flats may grow laterally as well as vertically, but vertical accretion seems to dominate. In many areas lateral accretion is limited by tidal channels and coastal erosion. Since environmental conditions on tidal flats are of a very specific nature, with alternatingly wet and dry periods, strong currents, temperature fluctuations, etc., they are inhabited by a typical fauna of intertidal animals. Most of these live buried in the sediment, or in protective attached shells. Diversity in tidal flat ecosystems is low, but biomass and production can be very high, due to the input of organic matter from the sea. This in turn attracts →carnivores, which exhibit special adaptations to life in tidal areas, such as e.g., tidal activity and feeding cycles, and tidal migrations. See also →zonation.

tidal marsh see →salt marsh.

tidal pools (syn. rock pools) holes or crevices in the intertidal zone of rocky seashores, where during low tide seawater is left behind. Such pools have a balanced population of plants and animals, and can be seen as a →microcosm of the sea. Due to evaporation and rainfall, rock pools show higher variations in temperature, →salinity, and →pH than the sea; the highest pools undergo the largest fluctuations and consequently have the lowest number of species. As "natural aquaria" rock pools offer an easy opportunity to study intertidal organisms →in situ.

tidal prism the amount of water in a small, semi-enclosed area between the high and the low water level. Assuming that high and low water occurs everywhere at the same time, it is equivalent to the in- and out-flowing tidal water mass.

tidal range see →co-range lines.

tidal wave often the term "tidal wave" is used without direct connection with the →tides, for destructive wave phenomena such as →storm surges or →tsunamis. However, in tidal theory the tidal wave is the result of the dynamic response of the sea to the periodically varying tidal potential. A tidal wave thus has the character of a →long wave in the ocean. In the deep open ocean its effect is usually small, but in more enclosed and shallow regions the response to the open-ocean tides may result in more or less complete resonance, giving strong tidal variations. For certain regions the tidal wave can be described as a →Kelvin wave, a →Sverdrup wave or a →Poincaré wave.

tidal zone (syn. →litteral zone) see →zonation.

tide periodic variation of sea level and associated periodic currents caused by the gravitational attraction between moon and earth (→lunar tides) and sun and earth (→solar tides). In such a system of two celestial bodies rotating around a common center, a field of gravitational and centrifugal forces is acting that modifies the earth's own gravitational field. This field of tide-generating forces is symmetrical: it is directed to the sublunar (subsolar) point and the opposite (antipodes or nadir) point. By the diurnal rotation of the earth around its axis in this field each point on earth experiences a regular variation of its →geopotential field. This variation (the tide-generating potential) acting on the oceans causes a variation of the →sea level. In more detail, astronomical variations in this system cause further variations at different periods (see →astronomical tides). The actual tide at any place and moment is the result of their contributions (see →harmonic analysis). The resulting tide is not directly related to the phase and amplitude of the local tidal potential, because of the dynamic response of the seas to different periods.

tide-generating potential see →tide.

time scales applied to water motions or transport problems, serve to assess the relevance of the various possible effects or processes for a given situation. Applied to dynamics the time scale usually appears as a "typical" time in the assessment of a nondimensional number. Applied to the transport of water or dissolved substances through a sea area several concepts such as age or transit, flushing, and residence time are used, but often differently defined. Usually the transit time of an element (e.g., water particle) is the time it remains in the area before leaving again. Its age is the time elapsed since an element entered for the last time. →Residence and flushing time are sometimes synonymous with one of the foregoing concepts, but different definitions are encountered. Theory in relation to these time scales usually presupposes stationary conditions. In that case the average transit time is the quotient of the total mass present and the total flux per unit of time. This quotient is sometimes called the →turnover time. See also →chronostratigraphy and →geological time scale.

Tintinnina (tintinnids) suborder of the phylum →Ciliophora, ciliate unicellular organisms (± 50 µm) living in the plankton. They differ from other ciliates by the possession of a vase-like skeleton or lorica, which sometimes consists of particles collected from the surrounding water and cemented together, as, e.g., →coccoliths. They may form an important link in the food-chain between →pico- and →nanoplankton. See also →Ciliophora.

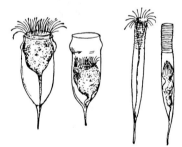

Tintinnina. (Hardy 1958)

tombolo sand bar connecting an island with the mainland.

topographic wave see →trapped wave.

tortuosity structural characteristic of a sediment that describes the curvature of the pore space. The tortuosity (q) of a sediment is defined as the ratio of the actual distance (l_a) a solute or water parcel must travel between two locations and the length (l) of a line connecting those locations: $q = l_a/l$. Some authors, however, define tortuosity as the squared ratio, $(l_a/l)^2$.

toxic acting as, or having the effect of, a →poison or →toxin.

toxic organisms see →venomous marine organisms.

toxicity the quality, relative degree or specific degree, of being toxic or poisonous. Acute toxicity: the consequences of "high" doses of a →toxin within a short time. As a quantitative measure the LD_{50} (lethal dose) or LC_{50} (lethal concentration) is used: the calculated dosis or concentration causing 50 % →mortality in an experimental population. Semi-chronic toxicity: the effect of generally low or sublethal concentrations of a toxin during an exposure time of about 10 % of the life time of an experimental organism. Chronic toxicity: the consequences of sublethal concentrations of a toxin to physiological mechanisms in organisms exposed during their complete life cycle (or 80 to 90 % of it).

toxins any of a group of poisonous, usually unstable, compounds generated by microorganisms, plants, animals, or of anthropogenic origin (e.g., industrially synthesized).

trace elements chemical elements occurring in seawater in concentrations typically below the micromolar level. Virtually all metals, largely the transition metals, occur at such nanomolar to picomolar trace levels. Some, such as Fe, Mn, Cu, Zn, Mo, V and Co, are micronutrients which occur in sufficient abundance for algal growth in the sea.

trace fossil see →ichnofossil.

trace metals see →trace elements.

tracer substance or property that marks water or sediment at a given location and that can be followed during further transport. Such substances may be artificially introduced like some →radionuclides (radioactive tracers) or dyes (rhodamine), or they may occur naturally (e.g., some isotopes). Radioactive tracers are also used in measuring biological processes in the sea, e.g., →^{14}C, since its introduction by E. Steemann Nielsen in 1952, in measuring →primary production. Properties that can be used as tracers are →potential temperature, →salinity and potential →vorticity. A →water mass with well-defined properties is often used to trace large-scale circulation in the ocean. Tracers that are nonconservative (nutrients or unstable isotopes) are less suitable for studying long-term transport, but are sometimes helpful in combination with conservative substances, to establish the transport time. Anthropogenic substances recently introduced into the environment and detectable at very low concentrations are used more and more as tracers. As the time of introduction into the environment is often known, they can also be used to determine the "age" of the water. Examples are dissolved Si as water-mass tracer, Sr and Cs radioisotopes as tracers of artificial inputs by mankind, Pb isotopes as tracers for origin of gasoline-derived Pb in the oceans, $^{3}H/^{3}He$ as tracer for bomb testing inputs, chlorofluorocarbons derived from anthropogenic inputs as tracers for downmixing of surface waters. See also →cosmogenic nuclides and →radioisotope.

trade winds winds on the equatorial side of the high-pressure cells in the subtropics. They are persistent in direction (NE in the northern, SE in the southern hemisphere). See also →climate.

transform fault fault with horizontal displacement only, marking a →plate boundary and terminating abruptly, at each end, at another plate boundary. Transform faults have different names depending on the features they connect. Most common are transform faults offsetting the axis of →mid-ocean ridges (R-R transform faults). See also →plate tectonics.

transform plate boundary see →plate boundary.

transgression spread or extension of the sea over a land area, and the consequent evidence of such an advance, e.g., marine sediments deposited on former land surfaces. Transgressions may be worldwide as the result of global sea-level rise or locally caused by the subsidence of land. Compare →coast and →regression.

transparency see →optics in the sea.

trapped wave due to a spatial variation of the phase velocity the large-scale propagation of →free waves can be limited by →refraction to a narrow range of directions called a wave guide. A sound channel (see →acoustics) can act as a wave guide for acoustic waves. Similar effects can occur for gravity waves, such as →edge waves, double →Kelvin waves and bottom-trapped waves, trapped by the depth-dependent variation of the phase speed. These waves are also called topographic waves. →Internal waves can become trapped near the →thermocline, due to the depth-dependent variation of the →Brunt-Väisälä frequency. →Equatorial waves are trapped because of the change of sign of the →Coriolis force at the equator.

traveling wave see →progressive wave.

trawl net towed by a motor vessel through the water (pelagic trawl) or over the seabed (bottom trawl), consisting of a conical bag of heavy netting and a construction to keep the net entrance (or mouth) open. Pelagic trawls are mounted with "doors" (rectangular boards that, when pulled through the water are driven apart to keep the entrance open); bottom trawls are kept open either by weighted doors that run over the seabed (otter trawl) or by a horizontal beam with sledges on both ends (beam trawl). The catch collects in the rear (cod end) of the trawl, which can be opened from behind when on deck after a haul.

trench relatively narrow, elongate depression of the deep-sea floor, with steep sides and oriented parallel to the trend of the continental edge. A trench is about 2 km deeper than the surrounding sea floor and may be thousands of kilometers long. In →plate tectonic context, often a discontinuity of the →lithosphere where one plate plunges beneath another. Commonly associated with a volcanic arc or →island arc in the overriding plate displaced from the trench. See →sea-floor topography.

Triassic see →geological time scale.

Trilobita (trilobites) subphylum of the phylum →Arthropoda. Extinct, exclusively →paleozoic marine bottom dwellers. More than 1500 genera have been described. Probably related to ancestors of →Crustacea.

triple junction in →plate tectonics, a point or area where three →plates meet.

tritium (T) cosmogenic →radioisotope of hydrogen with mass = 3, as opposed to mass 1 for normal hydrogen and mass 2 for stable isotope deuterium. Additional input to ocean surface largely through bomb testing in the early 1960s, as ^3H_2O. Radioactive decay leads to formation of ^3He. The ensuing shift in $^3He/^4He$ isotopic ratio in seawater serves as time clock for the age of the water mass.

trochophore (syn. trochosphere) free-swimming larval stage of, e.g., →Polychaeta, →Bryozoa, →Brachiopoda, and →Mollusca.

trochosphere see →trochophore.

trophic level level in a food chain at which an organism takes its food. When primary →producers are alotted trophic level 1, those eating plants (→herbivores) stand one step higher in the food chain, at trophic level 2; proceeding stepwise, →carnivores eating herbivores stand at level 3, top carnivores at 4, etc. Because at every trophic step in the food chain the food is only partially converted into edible new biomass, the organisms feeding at successive trophic levels in a system together build a food pyramid with a large mass of plant food eaten at the basis, and little meat eaten at the top by the highest carnivore level. In reality, most organisms take their food from more than one trophic level, so that the concept is only a rough approximation; thus trophic levels become average trophic levels for a defined group of animals. For instance, all North Sea fish together stand at an average trophic level of 3.7. Trophic structure of a given →ecosystem shows the stepwise conversion of an amount of plant material produced at the basis, in the different trophic levels. The trophic structure can give only a simplified view of a system, but it can be highly informative in comparing different systems. See also →food chain.

trophic structure see →trophic level.

tropical area/zone basically the part of the world between the tropics (Cancer and Capricorn), which lie at the latitudes 23°27′ north and south. In the marine environment, due to the prevailing ocean currents, the tropical areas show a more irregular distribution, related to high temperatures of the surface water and the occurrence of tropical organisms such as reef-building corals and small brightly colored fish. See also →biogeography.

tropical cyclones relatively small but very deep atmospheric disturbances that occur in the subtropics and may cause heavy damage in coastal areas. When they move over the sea there is a strong interaction with the surface waters:

the latent heat released when water evaporated from the ocean condenses again at high altitude is a major factor in lowering the atmospheric pressure. They are also known under the name of hurricane (America) or taifun (Japan).

trough see →trench.

tsunami a long-period wave with the character of a →long wave generated by a submarine earthquake. A tsunami can be observed at large distances from the epicentre and its amplitude may increase due to local topographic factors, resulting in considerable damage and loss of life when reaching a coast. When the epicenter and time of occurrence of an earthquake is known, the time of arrival of a tsunami can be predicted because the wave speed of shallow water waves is a function of depth. On this basis warning systems have been developed in the Pacific, where tsunamis mainly occur.

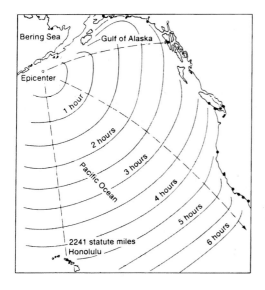

Tsunami. (After Pipkin et al. 1977)

Tunicata (tunicates) see →Urochordata.

Turbellaria class of the phylum →Platyhelminthes or flatworms. Marine, oval to elongated unsegmented worms, mostly bottom dwellers and members of the →meiofauna. The flattened body is covered with →cilia, and contains muscle layers with gland cells providing the mucus for adhesion. The head often has lateral projections or →auriculae. The intestine is a blind sac, the

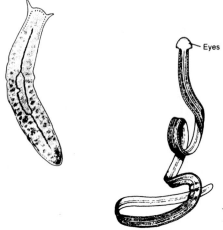

Turbellaria. (From Barnes
1987 after Hyman 1951)

mouth is for both ingestion and egestion. They feed on dead animals or are
predators feeding on small invertebrates, some on algae.

turbidite deposit resulting from a (high-density, high-velocity) →turbidity
current. The model for the ideal single turbidite sequence (the Bouma
sequence) is made up of five units with specific sedimentary structures.
From bottom to top these are: (a) graded interval (see →graded bedding);
(b) lower interval of parallel lamination; (c) interval of current ripple
lamination; (d) upper interval of parallel lamination; (e) pelitic interval.
Many turbidites, however, lack one or more of these intervals: as flow
velocity of the turbidity current wanes and the distance across the basin
increases, less and less coarse material is found in the deposits. (Figure see p.
289)

turbidity the degree to which water contains particles that cause backscattering
and extinction of light (see →optics). High turbidity is caused by a high
content of silt or, sometimes, of organic particles.

turbidity current short-lived, powerful, gravity-driven current consisting of a
dilute mixture of sediment and water of a density greater than the
surrounding water; the sediment is supported by the turbulence of the water
within the flow. Two types are distinguished: low velocity (low-density
turbidity currents), and high velocity (high-density currents). Turbidity
currents may be triggered by various events, such as earthquakes, storm
waves, over-supply of sediment, and have a flow route starting on the
continental shelf, continuing downslope through a →submarine canyon

	Grain Size		Bouma (1962) Divisions	Interpretation
	Mud	E	Interturbidite (generally shale)	Pelagic sedimentation or fine grained, low density turbidity current deposition
	Sand–Silt	D	Upper parallel laminae	? ? ?
	Sand–Silt	C	Ripples, wavy or convoluted laminae	Lower part of Lower Flow Regime
	Sand–Silt	B	Plane parallel laminae	Upper Flow Regime Plane Bed
	Sand (to granule at base)	A	Massive, graded	? Upper Flow Regime Rapid deposition and Quick bed (?)

Turbidite. (Reineck and Singh 1973). Ideal sequence of sedimentary structures in a single turbidite

and ending (with deposition of the sediment as a →deep-sea fan) on the continental rise (see →sea-floor topography and →turbidite).

turbidity maximum phenomenon characteristic of many →estuaries where freshwater mixes with seawater. Due to the difference in →density of freshwater and seawater, a countercurrent of seawater occurs underneath the outflowing freshwater from river through estuary to the sea. Settling sedimentary particles from the freshwater layer are retransported up the estuary by the bottom countercurrent, causing an accumulation of suspended matter in the low brackish zone of the estuary. This is called the turbidity-maximum zone.

turbulence irregular motion of a fluid superimposed upon other, more regular, currents, and by which transport of momentum, heat and dissolved substances may occur, in addition to the transport by the regular current. The transport of momentum gives an internal or boundary-layer friction that is comparable to viscous friction, but that is much larger and therefore often considered the sole friction force in the equations of motion. Oceanic motions can nearly always be considered turbulent. Turbulent mixing of heat and dissolved substances (eddy diffusion, see →eddy) is also much more important in oceanographic practice than molecular diffusion.

turnover rate presents the fraction of the amount of a substance in a component which is replaced over a certain time span. The biomass turnover rate or relative production (P/B ratio, where P = production and B = biomass) is a useful measure of the rate of functioning of a certain biomass in an →ecosystem. Turnover time is the reciprocal of the rate, i.e., the time needed to replace a quantity of substance equal to the amount in the component.

turnover time see →turnover rate and →time scales.

turtles, marine see →Reptilia.

U

ultraviolet radiation see →UV light.

UNCLOS I, II and III United Nations Conferences on the →Law of the Sea (LOS), which took place in 1958 (I) in Geneva, 1960 (II) also in Geneva and 1967–1982 (III) in New York-Caracas-Montego Bay. These conferences established an internationally accepted Law of the Sea.

upwelling upward movement of deeper water under the influence of →divergence of water at the surface, usually caused by differences in →Ekman transport. The rise of nutrient-rich deep water into the →photic zone stimulates →phytoplankton growth and, subsequently, that of →zooplankton and →nekton. Upwelling, or →divergence, may be induced also topographically when currents pass islands, peninsulas or seamounts, and →eddies arise, but are especially induced by wind patterns driving surface waters away from the coast (→coastal upwelling. Major upwelling areas, with large fish and squid populations (and fisheries) along the west coasts of Mauritania and Peru are due to constant trade-winds blowing parallel to the coast.

uranium (U), a radioactive metallic element (used, e.g., in nuclear weapons and as fuel in nuclear power stations) with the atomic weight is 238.03. The uranium →isotopes ^{235}U and ^{238}U comprise 0.715 and 99.28 % of natural uranium, respectively. The concentration of uranium in seawater may be different in various marine systems (estuaries, coastal seas, oceans), the average values for oceanic waters around an overall mean of 3.3 μg l^{-1}, which correspond to a radioactivity of 2.2 pCi l^{-1}. In the ocean ^{234}U and ^{238}U occur in dissolved form and are mother isotopes of a family of radioactive daughters. The radioactive decay series of ^{238}U (^{234}U) provide important time clocks for the rates of oceanic processes.

Urochordata (syn. Tunicata) subphylum of the phylum →Chordata. The Chordata are divided into the subphylum →Vertebrata, and the two subphyla (Urochordata and →Cephalochordata) that lack the characteristic backbone, but during any stage of their embryonic development possess properties that cause them to belong to the Chordata: a notochord, a hollow dorsal nerve cord, and pharyngeal clefts. After metamorphosis the notocord is lost and the nerve cord degenerates. See further →Ascidiacea and →Thaliacea.

UV light ultraviolet light, the part of the electromagnetic spectrum with wavelengths of 200 to 400 nm. Depending on the physiological effects, ultraviolet light is classified as UV-A (320–400 nm), UV-B (280–320 nm) and C (200–280 nm). The shorter wavelengths, with higher energy, are more harmful to living organisms. Ultraviolet light, especially the wavelengths shorter than 290 nm, is absorbed and reflected in the atmosphere. The ozone layer in the stratosphere is responsible for the absorption of short-wave UV light; it is therefore of fundamental importance to life on earth. Due to human activities, the ozone layer above the polar region is gradually eroding (ozone holes) and more and more UV-B radiation reaches the earth's surface.

V

vacuole fluid- or gas-filled space within a cell.

vagon all benthic organisms which move about; opposite of →seston.

Väisälä frequency see →Brunt-Väisälä frequency.

validation see →models.

Van Veen bottom sampler (syn. Van Veen grab) see →bottom samplers.

variance the mean square of the deviation of a variable quantity from its mean value. See also →statistics.

varve thin bed of sediment or lamina or sequence of laminae deposited in a single year. The well-known glacial varve consists of a thin pair of graded layers yearly deposited in glacial lakes; the lower "summer" layer is made up of relatively coarse-grained sediment transported by meltwater streams, whereas the upper "winter" layer consists of very fine-grained (clayey) sediment slowly deposited from suspension in quiet water when the streams were ice-bound. Marine varves, the three rings of the oceans, contain dark laminae originating from seasonally cyclic biogenic production or seasonal changes in →terrigenous sediment supply. Examples are →diatom varves in the Guyana Basin, Gulf of California, and the varves in the Santa Barbara Basin, California, caused by increased sedimentation during the winter rain period on land. Varves can be counted to measure the age of the deposits involved. See also →graded bedding.

veliger larva of →Mollusca, which develops from a →trochophore. Ciliated bands have become larger; foot, mantle, shell, and other organs of adult are present.

velum contractile ectodermic fold extending from the edge of the body of a →medusa.

venom →poison secreted by certain animals (e.g., snakes, fish), usually injected by bite or sting.

venomous marine organisms animals which have the ability to administer a zootoxin by means of a venom apparatus. This distinguishes them from organisms which are →toxic after they are eaten. Such toxins are called →poisons, although the term poisoning is sometimes used in the general sense, referring to reactions to both oral poisons and venoms. Some examples of venom apparatuses are →nematocysts (e.g., in the →scypho-meduse *Chironex fleckeri*, the sea wasp), venom darts (in →gastropods of the *Conus* family), salivary glands and beak of →cephalopods, spines (e.g., in *Synanceia*, the stonefish and *Trachinus*, the weever) and teeth (in sea snakes). Venoms are mostly used to paralyze prey, but they may also serve as defence.

vent see →hydrothermal vents.

ventral situated at, or relatively nearer to, the side of an animal which is normally directed downwards; opposite of →dorsal.

Vertebrata (vertebrates) subphylum of the phylum →Chordata, constitute 5 % of the species of the animal kingdom, characterized by an endoskeleton that includes a backbone consisting of a series of vertebrae. The vertebrae develop around the notochord (see →Chordata), in most vertebrates only present during the embryonic development. The Vertebrata contain the classes Agnatha (jawless fish), Placodermi (fossil armoured fish), Chondrichthyes (cartilaginous fish), Osteichthyes (bony fish), Amphibia (frogs, toads, salamanders), Reptilia (turtles, crocodiles, snakes and lizards), Aves (birds) and Mammalia (mammals). Representatives of all classes, except Amphibia, are found in marine biota.

vertical distribution see →distribution.

vertical migration one of the most striking aspects of zooplankton behavior is the diurnal vertical migration exerted by many species of planktonic herbivores (comprising members of many phyla) and carnivores, including pelagic fish. With echosounders, but also with plankton nets that can sample different depth layers, it has been demonstrated that planktonic animals move upwards over distances of sometimes hundreds of meters at sunset, to spend the night in surface layers. At sunrise, the opposite vertical movement brings them back to the depth, where they stay during the day (see →scattering layer). These vertical movements must require considerable energy, and must therefore have a clear function. However, it has not yet been possible to determine which of a number of hypothetical explanations is the correct one, though the changes in light intensity in the morning and evening are generally accepted as the stimulus governing this behavior. Explanations suggested are, for instance, that the animals can reduce their energy requirements by staying part of the day in colder water layers in the depth, or that vertical migration leads to an optimal use of phytoplankton resources which can grow during the day unhampered by grazing. It is often thought that evading visually hunting predators in surface waters during the day might be a functional reason as well.

vertical mixing of water is caused by wind at the surface, by →shear at the bottom or between different layers, and by convection at the surface. In shallow waters, wind and (tidal) current are sufficient to keep the water column well mixed, i.e., homogeneous in temperature and salinity. In deep waters, a wind-induced mixed layer is found over only a fraction of the total depth, the remainder being stably stratified. In temperate regions, surface cooling in autumn and winter renders the surface layer gravitationally unstable leading to convective overturning, which deepens the upper mixed layer, down to the permanent →thermocline.

vestigial organ organ, of which the size, structure and function have diminished and simplified in the course of evolution until only a trace remains. Such vestiges may still have an important function in organizing embryonic development, even if their adult function has disappeared.

viable count is the estimation of the number of live cells in a microbial population and thus differing from the total count, i.e., the total amount of bacteria seen by microscopic methods, in which also dead cells are included.

vibraculum (pl. vibracula) a modified →zooid in →cheilostomatous Bryo-zoa; the →operculum of the zooid is modified into a long bristle, which can be moved in one plane and is used to sweep away particles settling on the colony. In some free-living cheilostomes they can also be used in locomotion of the entire colony by coordinated movements of all the vibracula together.

viroids small pieces of →RNA which are not complexed with protein and have →virus-like properties.

virus a genetic element containing either →DNA or →RNA that is able to alternate between intracellular and extracellular states, the latter being the infectious state. Viruses are pathogenic agents, parasitic in bacteria, plants, and animals, unable to multiply outside the host. They are so small ($< 0.2 \mu m$) that they pass through filters which retain bacteria.

visceral mass in soft-bodied animals the ventral organs (intestine, stomach, digestive gland, heart, kidneys) are coalesced into the visceral mass.

viscosity measures the ability of moleculair transport to eliminate the nonuniformity of fluid velocity. The molecular transport of momentum in a fluid depends on the composition of the fluid and its temperature, and is expressed by its viscosity. The dynamic viscosity gives the relation between the internal friction per unit surface and the current →shear; the kinematic viscosity is the dynamic viscosity divided by the density of the fluid.

vitamins are accessory food constituents which are necessary only in minute amounts for a normal functioning of organisms. Absence or shortage leads to deficiency diseases. They are part of biologically active substances, e.g., vitamin B_{12} is primarily released by bacteria and enhances the development of some phytoplankton species.

volcanic arc volcanic belt, commonly associated with an →island arc or active →continental margin.

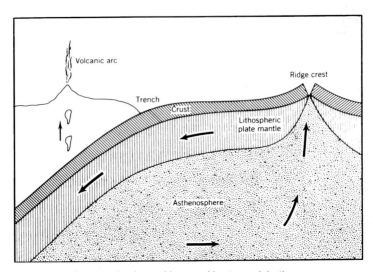

Volcanic arc. Cross-section from mid-ocean ridge to a subduction zone

volcanism the whole of processes associated with the transfer of →magma from the interior to the surface of the earth. If magma reaches the surface (then called lava) it builds features of various sizes and shapes. Most common are the →mid-ocean ridges where volcanism accounts for the accretion of new oceanic floor. Next most important are the great →basalt plateaus that occur within →plates (Hawaii) or slowly opening spreading centers (Iceland). This volcanism is characterized by regular quiet eruptions of fluid lava from several fissures or vents. In places, lava flows accumulate to thicknesses of more than 2 km. The great subsurface extent and the flat-topped plateau form are due to the relative fluidity of the lava, whereas rapidly chilling by seawater results in the steep-sloped sides of e.g., →seamounts and →tablemounts. The classic individual →volcanoes (e.g., Fuji, Vesuvius) are characterized by relatively viscous lava confined in a crater. Beneath the rapidly solidifying lava, gas accumulates and builds up pressure, resulting in sometimes explosive emissions of gases, lava and pyroclastic materials (→tephra), tending to build curved sloped cinder cones around the central vent.

volcano (1) vent in the earth's crust, connected by a conduit to the earth's interior from which issue molten lava, pyroclastic materials and volcanic gases; (2) mountain built up by the materials ejected from the interior of the earth through a vent. See also →volcanism.

Volvocida order of the subphylum →Mastigophora (phylum →Sarcomasti-gophora, see →Protozoa) of mainly freshwater species. Regarded by botanists as the order Volvocales of the class →Chloropyta or green algae. The cells have a cellulose wall and can be nonmotile or motile (e.g., *Chlamydomonas*) and colony forming (e.g., *Volvox*).

vortex the rotating motion of a large number of material particles around a common center. A line at every point parallel to the local →vorticity vector (the axis of rotation) is called a vortex line in →hydrodynamics. The vortex lines drawn through a closed contour form a vortex tube, and the fluid that is contained within this tube is a vortex filament, or briefly, a vortex. In hydrodynamics, vortices are convenient in describing the changes in →vorticity.

vorticity of a fluid element is a measure of the spin of that element, proportional to its angular momentum. Positive vorticity means a clockwise rotation, negative an anticlockwise rotation. In vector analysis vorticity is defined as the curl of the velocity. On the rotating earth absolute vorticity is the sum of the vorticity relative to the earth (relative vorticity) and the rotation of the earth (planetary vorticity). Potential vorticity is a quantity that in homogeneous water is equal to absolute vorticity, divided by depth. In stratified water it depends on the stratification. This potential vorticity is a conservative quantity, and is sometimes used as a dynamic →tracer.

W

warm-blooded animal see →homeotherm.

waste economically useless products, resulting from almost any human activity, that are disposed of in the environment. Domestic waste is waste introduced, in the environment by municipal communities, and individual housekeepings. Industrial waste is waste introduced into the environment by industrial activities and can be of different kinds: side products of any synthetic substance; acid-iron wastes [either sulfuric acid (H_2SO_4) or hydrochloric acid (HCl) or a combination of these two]; pharmaceutical waste; fly ash; consolidated fly ash and scrubber →sludge. Several types of waste are dumped into the ocean; in each type, there are large variations or ranges in physical and chemical properties. Dispersion, advection, and settling of a waste in the ocean will depend on physical properties such as bulk density, solid density, suspended solid concentration (for a liquid), and particle size. Classification of wastes may be based on their origin (production, distribution, and consumption waste), their physical conditions (solid, liquid, slurry, smoke, or gas), the way of possible treatment (combustible, inert, compostable, recyclable) or other specific characteristics (toxic, pathogenic, and radioactive waste). See also →ocean dumping.

waste disposal see →ocean dumping.

water balance of the ocean the balance of all incoming and outgoing water at the order of $450 \cdot 10^3$ km^3 y^{-1}, via →air-sea exchange, via land (run-off) or by formation and melting of →ice. On average, the water balance of the ocean is assumed to be in equilibrium. However, at very long time scales there can be a residual term that may lead to sea-level changes (see →climatic variation).

water gun see →seismic instruments.

water mass the term water mass is often used in a specific sense, indicating water with a well-defined combination of properties such as →potential temperature and →salinity, but sometimes also chemical, optical, or biological properties, by which it can be discriminated as a special body of water (see →indicator species). Such a combination of properties is supposed to have been acquired in a certain area, and the idea is that the water mass can be traced back to its point of origin (source area). It is essential that the properties are conservative, i.e., that they should not be modified underway otherwise than by mixing between water masses. Therefore →potential temperature is to be preferred above actual temperature. In the analysis of water masses the →temperature-salinity (or T-S) diagrams are used. Homogeneous water masses are represented by a single point in such a T-S diagram, and are labeled by names indicating their source and level of occurrence. In such diagrams combinations of temperature and salinity measured at successive depths typically lie along curves that can be considered mixtures of two or more water types.

water quality an arbitrary term for the degree of →contamination of water.

watersprout wind vortex with the shape of a funnel reaching the water surface from a Cumulonimbus (thunderstorm) cloud, and pulling water from the water surface. Usually its radius is only some 100 m large. A watersprout results from strong atmospheric instability.

wave commonly used in oceanography for the small-wavelength surface →gravity waves, but physically also used for other wave phenomena. Waves can be divided into transversal waves (where the disturbance is at right angles to the direction of propagation) such as sea waves and large-wavelength tides, and into longitudinal waves (where the disturbance is in the direction of propagation). To this last category belong the acoustic waves, where the compressibility of the water is important. Mathematically the waves are described by a set of →wave characteristics. See also →seismic wave.

wave approach waves approaching the coast enter shallower water (shoaling). At an abrupt change in depth, partial reflection in waves may occur. Gradual decrease in depth causes →refraction of obliquely approaching waves. The →wave characteristics will change when depth decreases. After a first slight decrease in wave height closer to the coast the height and steepness of the waves increase until they become unstable and break. Empirically it has been found that this occurs at a depth of 1.2 times the wave height. The action of the waves after breaking is called the **surf**. By inertia the water runs up the shoreline (**swash**) and returns back along the slope (**backwash**).

wave characteristics describe a wave mathematically. Apart from the wavelength L, the amplitude A and the period T, the wave number $k = 2\pi/L$ and the angular frequency $\omega = 2\pi/T$ are used. Individual →progressive gravity waves move with a phase speed $c = \omega/k = L/T$. Generally, this speed is a function of depth and wave characteristics, but for relatively deep water it becomes depth-independent (see →short waves). For relatively shallow water (see →long waves) the phase speed is only a function of depth (D) and gravity (g): $c = \sqrt{(gD)}$. As →wave energy propagates with a speed that depends on phase speed, most gravity waves, except long waves, have a wave number-dependent propagation, and are subject to →dispersion. See also →wind waves.

wave energy in waves energy is continuously changing from one form to another and vice versa. In →gravity waves at the sea surface or at depth (→internal waves) there is an exchange between the kinetic energy of the

→orbital motion and the potential energy of the sea surface. It is usually expressed as energy density, i.e., energy per surface area. This is proportional to the square of the wave height. Wave energy is transported by →**progressive waves** at a speed (the group velocity) that, apart from →**long waves**, differs from the speed of propagation of the individual waves (phase velocity). In →**short waves** the group velocity is half the phase velocity.

wave generation see →wind waves.

wave-induced current in surface waves water particles gradually move in the direction of wave propagation (see →Stokes drift). If the waves approach the coast this net water transport causes an equilibrium state with an elevation of the surface balancing the inward flow. In the case of an oblique →wave approach a longshore current develops between the breaker zone and the shoreline. The return of the inshore water transport by the waves occurs at distinct places by rip currents at topographic lows in the near-shore sand bars.

wave number see →wave characteristics.

wave recorder measuring waves is determining the variation in time of the sea level and eliminating long-period variations such as tides. Techniques used are electrical (short-circuiting a series of contacts by the water), underwater pressure gauges, and measuring the vertical displacement of a float or the vertical acceleration of a drifting buoy. In general these methods measure waves irrespective of their direction of propagation. Directional wave recorders require additional observations. Other types of wave measurements are →remote sensing of a wave field (stereo photography or radar measurements).

wave ripple see →ripple.

Wegener, Alfred see →continental displacement.

well logging measurement of physical properties of rock sections encountered in a bore hole (well). The measurements are made by a sonde lowered into the bore hole by a multi conductor cable. Usually several different measurements are made simultaneously during one withdraw such as electrical, acoustic and radioactive properties. The recordings are plotted as a continuous function on a common depth scale. The word log is used for the individual curves as well as the gathering of curves. Well logs are commonly referred to by type, e.g., gamma-ray log, which is the log of the intensity of the natural gamma radiation.

westerlies climatic zone, see →climate.

western intensification strengthening of current at the west side of the large oceanic →gyres. See also →boundary current.

wetlands areas of marsh, fen, peatland, or water; either natural or artificial, permanent or temporary, with water that is static or flowing, fresh, brackish, or salt. They include areas of marine water where depth at low tide does not exceed 6 m (according the definition given in the International Wetland Convention). They support a characteristic flora and fauna, especially waterfowl and constitute a resource of great economic, cultural, scientific, and recreational value.

whale fishery or rather whale hunting has a long history, and was originally restricted to coastal areas during the seasonal passing of migrating whale populations (in Europe: Norway, Portugal, and the north coasts of Spain). The animals were spotted on the sea surface during emergence, followed with small boats till they could be wounded and finally killed with hand-thrown harpoons, and hauled ashore. In the 17th century, especially the Dutch whale fishery increased rapidly, till an extensive whaling fleet hunted in the Northern Atlantic for the Greenland whale. Settlements and factories were set up on, e.g., Spitsbergen, to cook oil from the blubber and meat. Historical commanders' houses on the Frisian Islands, with their typical whale bone gates (jaws), remind us of the period when the Greenland whale was almost exterminated. When the whale fishery moved to northern England, and from there to the North American coast (Nantucket), hunt switched to the sperm whale, not only important for its oil and its ivory teeth, but especially for the →spermaceti: a wax-like material from the head, used to make fine oils for, e.g., the cosmetics industry. Ambergris, a solid grayish-black substance from the lower intestine of the sperm whale, formed around the beaks of squids eaten, and used by perfume makers, was also valuable. Due to the decrease in the sperm whale population at the end of the 19th century, whale fishery almost disappeared. With the invention of the harpoon gun, firing heavy explosive harpoons guiding steel cables, the hunt for the fast blue whale and fin whale started in the Antarctic seas in summer. Small and fast vessels (whale hunters) provided large factory ships with the inflated carcasses, which were cut up and cooked in the enormous oil kettles, though in some cases meat was frozen for human consumption (Japan). Due to the whale fishery, many species of (mainly) baleen whales have become almost extinct. Since 1931 there has been a sequence of international treaties aimed at the protection of the whales through regulation of the exploitation, and at present whale fisheries are closed worldwide except for a very limited catch for "scientific purposes".

whales see →marine mammals and →whale fishery.

white-capping breaking of →wind waves in the open sea.

wind flow of moving air. Near the surface of the earth the wind is influenced by friction (see →Ekman layer), at higher levels the wind is approximately →geostrophic. As the friction of the wind over sea is small compared with that over land, the wind is usually stronger over the sea. The wind shows variations in speed and direction at short time scales. In meteorological reports mean values are given; the variations around this mean constitute the atmospheric turbulence. The wind is measured with →anemometers. The conventional level is 10 m above the ground in an area free of obstacles. At sea these conditions are not easy to fulfill, as the measuring platform (ship or structure) disturbs the flow. The wind at sea is often described by an estimate of the wind force according to the →Beaufort scale (0 = calm, ..12 = hurricane). For each Beaufort number a range of wind speeds has been established empirically (Beaufort equivalents) by the World Meteorological Organization (WMO).

wind factor in empirical studies of the relation between wind and drift current one often finds a linear relation, with a so-called wind factor, as a proportionality coefficient. See also →drift current.

wind stress wind over sea exchanges momentum with the sea surface, partly by generating →wind waves. This process can be represented by a horizontal frictional force per unit of surface area of the sea, the wind stress. The wind stress depends on the square of the wind speed, the density of the air and a so-called →drag coefficient that is often taken as constant, but appears to depend on the atmospheric stability.

wind waves waves generated by wind blowing over the water surface. The exact mechanism of this process is not yet fully understood, but it is an important process in the →air-sea exchange of momentum. Empirical relations have been established, however, that allow the prediction of wave height and period from meteorological data. In such predictions the distance over which the wind blows in a uniform direction and at a constant speed (fetch) is an essential point. The fetch may be limited by the coast or by the size of the wind field itself. If the duration and fetch of the wind are long enough, a fully developed sea exists, in which gain and loss of momentum (dissipation, e.g., by breaking) are in equilibrium (see →wave characteristics).

Winkler titration a chemical determination of oxygen dissolved in water, developed by the German chemist L.W. Winkler in 1888. An alkaline manganese Mn^{2+} solution reacts very fast with oxygen to higher oxidized Mn compounds. Such compounds react in acid medium with iodide to form iodine. The yellow-brownish iodine solution is titrated with sodium thiosulfate to a faint yellow color. The addition of a small amount of starch solution changes the color to intense blue and the titration is continued until the solution is colorless. In classic descriptive oceanography this oxygen determination comes first after the characterization of a seawater sample by temperature and salinity determinations. A precision $>0.1\%$ can be obtained with a photometric end-point detection and this version opens the possibility to study open-sea →in situ diurnal changes for primary production studies and very low →BOD values. See also →oxygen demand and →oxygen.

X

Xanthophyceae yellow-green algae, class of the division →Heterokontophyta, created to categorize algae with green plastids, but otherwise resembling chrysophycean algae, especially in their flagellar apparatus (see →Chrysophyceae). Xanthophyceae contain →chlorophyll-c and heteroxanthin, but no fucoxanthin. Their storage products are fatty oils and chrosolaminarin, but not starch. Most of them are microscopic freshwater organisms. The genus *Vaucheria* forms threads of several cm. It also lives in coastal salt marshes. *Meringosphaera* is often present in tropical marine nanoplankton.

xanthophyll see →carotenoids.

XBT expendable bathythermograph, see →bathythermograph.

xenobiotic compounds completely synthetic chemical compounds not occurring naturally on earth.

xerophile adapted to growth in habitats where freshwater is rare, e.g., deserts, but also →salt marshes.

X-ray methods for digestion the rate of food digestion in fish can be followed by taking X-ray photographs of prey or specially treated food pellets in the stomach of the living fish at regular intervals.

Y

yeasts unicellular →fungi; cells may be spherical, oval or cylindrical. Cell division generally takes place by budding.

yellow substance see →Gelbstoff.

yield see →fisheries science.

yolk the reserve food for developing embryos in eggs. Yolk consists in dry weight of approximately 60 to 70 % protein and 30 to 20 % lipid. During development, yolk is transformed into embryo tissue with a conversion efficiency of about 60 to 90 %, depending on the rate of development and the amount of yolk utilized for metabolism of the embryo. In many fish species the hatching larvae still contain a considerable amount of yolk in the yolk sac, which is used by the larvae for further development and as an energy source during early life (yolk sac resorption).

yolk stage life stage during which larvae do not eat, but still live on the →yolk.

Z

Zoantharia see →Anthozoa.

zoea stage in the larval development of the →Malacostraca, which in most cases appears as the first free stage in the water, after the naupliar development (two or three stages) in the egg. In some species the zoea is formed during metamorphosis of the last of a number of free-swimming →nauplius stages. The abdomen of the zoea is already well developed, but the →thorax with a few pairs of appendages is still rudimentary in its hinder part. In crabs the zoea stage is followed by the first post-larval stage, the →megalopa.

zonation marine organisms often occur in belts or depth zones because of gradients in environmental factors. This is easily visible along rocky coasts, with a clear zonation from lichens high on the coast, and →barnacles and different algal zones lower down. Zonation is also present along sandy or muddy coasts, in →estuaries, and in the open sea. In the tidal or littoral zone, the part of the shoreline that is alternately exposed to air and wetted by submersion, splash or spray, can be discerned: the intertidal or eulittoral zone between high and low water; the supralittoral (syn. supratidal or surf zone), the splash zone or spray zone above the highwater line; and a sublittoral zone below low water and usually down to the shelfedge (ca. 200 m), sometimes divided into inner sublittoral (= inner shelf) and outer sublittoral (= outer shelf). The name surf zone is also used to denote the near-shore zone where both sea and swell waves peak up and break into surf, and change from waves of oscillation to waves of translation. The bottom of the ocean below the sublittoral zone is divided into a bathyal zone, roughly from the outer shelf down to ca. 2000 m (the 4 °C isotherm), an abyssal zone from roughly 2000 to 6000 m and a hadal zone from 6000 m down to the greatest depths, comprising all the deep trenches and characterized by a distinct fauna. The →pelagic environment of the water column can be equally divided into an epipelagic part (0–roughly 200 m), more or less equivalent to the →photic (= lighted) →zone; a mesopelagic (ca. 200–ca. 2000 m), a bathypelagic (ca. 2000–ca. 6000 m) and an abyssopelagic zone (>ca. 6000 m) forming together the →aphotic (= dark) zone. Not all authors agree on exact limits for the different depth zones, but the sequence as given here is generally accepted. Gradients in environmental factors effect the observed zonation in organisms. In the littoral zone particularly irradiation, exposure time, desiccation, grazing, and predation determine the zonation. Temperature is the main factor for

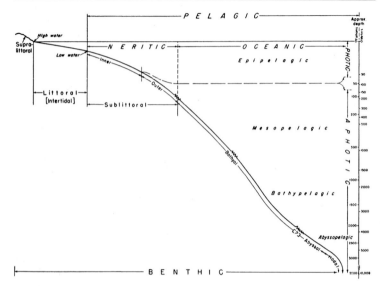

Zonation. (After Hedgpeth 1957)

zonation in deep water. The salinity gradient is the dominant ecological factor in zonation in estuaries. See also →hypsographic curve.

zoobenthos see →benthos.

zoogeography see →biogeography.

zooid individual →polyp in a colony. Often more than one type of polyps occur (polymorphism); they can be modified for feeding (gastrozooids), defense (dactylozooids), asexual reproduction, etc.

Zoomastigophora (syn. Zooflagellata) see →Mastigophora.

zooneuston animals living near or at the water surface of seas and lakes.

zooplankton animals in the →plankton. In marine zooplankton most animal phyla are represented, and both →herbivores and predators occur; herbivores feed on →phytoplankton and are in turn eaten by zooplankton predators. →Fecal pellets produced by larger zooplankton sink rapidly and produce food for →benthic organisms, but →coprophagy also occurs (particularly in copepods). Herbivorous feeding also releases nutrients which are used in so-called →regenerated primary production. →Copepods often dominate zooplankton.

zooplankton samplers although the smaller zooplankton, the microzooplankton, could be collected in sufficient quantities by taking water samples with the usual types of samplers, most of the zooplankton has too low a →density for this approach. For the quantitative study of the different groups of larger zooplankton, a variety of nets and samplers exists, which filter larger volumes through fine plankton gauze (i.e., nylon netting with meshes from 30 μm to several millimeters). Nets can be hauled vertically from bottom to surface or used horizontally in the current or towed behind a boat, with double oblique hauls. A current meter in the mouth opening provides an estimate of the volume of water filtered. →Closing devices allow the sampling of separate strata and are operated mechanically (by messengers; Juday net, Nansen net, Clark-Bumpus sampler), electrically (Gulf III high-speed sampler, various multinets, or nets with a rotating set of cod ends), or acoustically (Rectangular Midwater Trawl). Detailed vertical profiles of zooplankton are also made by large pumps (up to ca. 100 m depth) and by the Longhurst-Hardy recorder (see →continuous plankton recorder).

Zooxanthellae see →symbiosis.

zygote the fertilized ovum before cel division and differentiation.

References for Figures

Anonymous (1983) Annualk report of the continental Shelf Institute, Trondheim, Norway

Barnes R D (1987) Invertebrate zoology, 5th edn. Saunders Coll Publ, Philadelphia

Bascon W (1959) Ocean waves. W. H. Freeman & Co Beutelspacher H, van der Marel H W (1968) Atlas of electron microscopy of clay minerals and their admixtures. Elsevier Publ co, Amsterdam

Bjørnberg T K S (1972) Developmental stages of some tropical and subtropical planktonic marine copepods. In: Wagenaar Hummelinck P, van der Steen L J (eds) Studies on the fauna of Curacao and other Caribbean Islands, vol XL. Nat Studiekring v Suriname en de ned Antillen 69

Borradaile L A, Potts F A (1959) The Invertebrata. Cambridge Univ Press, London

Brock Th D, Madigan M T (1988) Biology of microorganisms. Prentice-Hall, Englewood Cliffs, N J

Brown F A (1950) Selected Invertebrate Types. John Wiley and Sons, NY

Cohen A C (1982) Synopsis and Classification of Living Organisms. Vol 2. McGraw-Hill Co. pp 191, 192

Dietrich G, Ulrich J (1968) Atlas zur Oceanographie. Bibliogr Inst Mannheim

Dietz R S, Holden J C (1970) The breakup of Pangaea. Sci Am Oct 223:30–41

Dobrin M B (1976) Introduction to geophysical prospecting, 3rd edn. McGraw-Hill, Tokyo

Duursma E K (1972) Geochemical aspects and applications of radionuclides in the sea. Oceanogr Mar Biol Ann Rev 10:137–223

Fairbridge R W (ed) (1966) The encyclopaedia of oceanography. Reinhold, Washington, DC

Fenchel T (1969) The ecology of marine microbenthos. Ophelia 6:1–182

Giesbrecht W (1982) Fauna and Flora Golfes Neapel. Monogr 19:1–831

Gill E A (1982) Atmosphere-ocean dynamics. Acad Press, New York

Hardy a (1958) The open sea. Its natural history: Part I. The world of plankton. Collins, London

Hedgpeth J W (1957) Classification of marine environments. In: Hedgpeth J W (ed) Treatise on marine ecology and paleoecology, vol I. Ecology. memoir Geological Society of America 67

Hersey J B (1963) Continuous reflection profiling. In: Hill M N (ed) The sea, vol 3. The earth beneath the sea. Interscience Publ, New York

Hesse H, Doflein F (1914) Tierbau und Tierleben, vol 11. Das Tier als Glied des Naturganzen. B G Teubner, Leipzig

Higgins R P (1971) A historical overview of kinorhynch research. In: Hullings N C (Ed.) Proceedings of the first international conference on meiofauna. Smithsonian Contributions to Zoology 76:25–31

Hoek C van den (1984) Algen. Thieme Verlag, Stuttgart

Hofker J (1977) The foraminifera of Dutch tidal flats and salt marshes. Neth J Sea Res 77:223–296

Holmes A (1965) Principles of Physical Geology. Nelson & Sons, London

Holstein T, Tardent P (1984) An ultra-high-speed analysis of exocytosis: nematocyst discharge. Science 223:830–833

Hovland M, Judd A G (1988) Seabed pockmarks and seepages. Graham & Trotman. London

Hyman L H (1940) The Invertebrates. Vol I. McGraw-Hill Book Co, NY

Hyman L H (1951) The Invertebrates. Vol II. McGraw-Hill Book Co, NY

Hyman L H (1951) The Invertebrates. Vol III. McGraw-Hill, NY

Hyman L H (1955) The Invertebrates. Vol IV. McGraw-Hill, NY

Käse R H, Clarke R A (1978) High frequency internal waves in the upper thermocline during GATE. Deep-Sea Res 25:815–825

Keeton W T (1972) Biological science, W. W. Naton & Co, New York

Kennett J P (1982) Marine geology. Prentice-Hall Inc, Englewood Cliffs N J

Knauss J A (1978) Introduction physical oceanography. Prentice-Hall, Englewood Cliff, N J

Landolt-Börnstein (1968) Oceanography. New Series vol 3, part c. Springer, Berlin Heidelberg New York

Lawrie W, Alvarez W (1981) One hunderd million years of geomagnetic polarity history. Geology 9 (9):395–396

Le Blond P H, Mysak L A (1978) Waves in the ocean. Elsevier Oceanographic Ser 20

Mclellan H J (1965) Elements of physical oceanography. Pergamon Press, Oxford

Mortimer C H (1977) Internal wave observed in Lake Ontario during the International Field year of the Great Lakes 1972. Spec Rep Univ Wisconsin-Milwaukee, Centre Great Lake Stud 32

Müller K (1982) Coastal research in the Gulf of Botnia. Junk, The Hague

Neshyba S (1987) Oceanography. Wiley & Sons

Neumann G, Pierson W J (1966) Principles of physical oceanography. Prentice-Hall, Englewood Cliffs, N J

Open University Oceanography Course Team (1989) The ocean basins: their structure and evolution. Pergamon Press, Oxford

Pritchard D W (1955) Estuarine circulation patterns. Proc Am Soc Civil Eng 81:717

Proudman S, Doodson A T (1924). The principal constituent of the tides of the North Sea. Philos Trans R Soc Lond A 224:105

Reimnitz E, Barnes P W, Tau Rho Alpha (1973) U. S Geological Survey Miscellaneous Field Studies. MF 532

Reineck H-E, Singh I B (1973) Depositional sedimentary environments. Springer Berlin Heidelberg New York

Reiswig H M (1975) The aquiferous systems of three marine Demospongiae. J Morphol 145 (4): 493–502

Riedl R (1963) Fauna und Flora der Adria. Verlag Paul Parey, Hamburg

Riedl R (1983) Fauna und Flora des mittelmeeres. Verlag Paul Parey, Hamburg

Ripkin B W, Gorsline D S, Casey R E, Hammond D E (1977) laboratory experiments in oceanography. Freeman San Francisco

Robertson L A (1988) Aerobic denitrification and heterotrophic nitrification in *Thiosphaera pantotropha* and other bacteria. Thesis, T. U Delft

Romer A S (1962) The vertebrate body. Saunders, Philadelphia

Scharnow U (ed) (1978) Grundlagen der Oceanologie. VEB Verlag für Verkehrswesen, Berlin

Schütt F (1895) Die Peridineen der Plankton Expedition. Ergebnisse der Plankton-Expedition des Humbold-Stiftung, Bd IV 1–27, 27 Tafel, Lipsius & Tischer, Kiel

Shepard F P (1963) Submarine geology, 3rd edn. Harper & Row, New York

Sverdrup H U, Johnson M W, Fleming R H (1942) The oceans. Prentice-Hall, Englewood Cliffs, N J

Tattersall W M, Tattersall O S (1951) The British Mysidacea.

Thompson T E (1976) Biology of opisthobranch molluscs, vol I. The Ray Society. BMNH, London

Tolmazin D (1985) Elements of dynamic oceanography. Allen & Unwin, Boston

Williams J (1973) Oceanographic instrumentation. Maval Institute Press, Annapolis, Maryland

Wimpenny R S (1953) The plaice. Arnold, London

Wyllie P J (1976) The way the earth works. Wiley & Sons Inc, New York

Younge C M (1975) Giant clams. Sci Am 232 (4):96–105